CONVERTING for FLEXIBLE PACKAGING

Adolph Miller, Ch.E.

TECHNOMIC PUBLISHING CO., INC.
LANCASTER · BASEL

Converting for Flexible Packaging
a **TECHNOMIC®** publication

Published in the Western Hemisphere by
Technomic Publishing Company, Inc.
851 New Holland Avenue, Box 3535
Lancaster, Pennsylvania 17604 U.S.A.

Distributed in the Rest of the World by
Technomic Publishing AG
Missionsstrasse 44
CH-4055 Basel, Switzerland

Printed in the United States of America
10 9 8 7 6 5 4 3 2

Main entry under title:
Converting for Flexible Packaging: A Primer

A Technomic Publishing Company book
Bibliography: p. 177
Includes index p. 179

Library of Congress Catalog Card No. 93-61005
ISBN No. 1-56676-061-5

To June

CONTENTS

PREFACE

THIS book is the first to cover all the processes that constitute converting—the technology of making a variety of material substrates usable for flexible packaging.

Converting has developed to a combination of high science and artistic craft, and both components have been taken into account in the text.

The intent of this book is threefold: to give the reader an appreciation for the converting process as a whole; to aid in the understanding of individual converting processes; and to communicate how converting has become the crucial technology it is today.

In addition to the main unit processes within the converting industry, the book also treats two ancillary, but critical, operations that affect converting: formulating and permeability. A chapter describes how to formulate, since the converter must be aware of the composition of the materials being printed, coated or laminated. A prime objective of converters is also to attain as low a permeability in a finished web as possible, a goal often attainable only through suppliers' data. The text furnishes information on the permeability of a wide range of packaging materials.

It is hoped that the reader finds the book both instructive and entertaining.

North Palm Beach, 1993

CHAPTER 1

Introduction and History

DEFINING AN INDUSTRY

THE industry doing converting for flexible packaging is little known, less understood and much taken for granted by the general population. The packaging of a product in plastic film or a metal foil composite produces no oohs and aahs.

It has become quite commonplace.

Yet the industry seems to draw more than its fair share of attention from state, federal, and local environmental agencies. It comes to public attention when it is denounced – with some lack of moderation – as a contributor to our waste problems. And while it struggles to conform to a multitude of regulations, some contradictory, many inconsistent and ever-changing, and others more than tinged with the color of politics, in the background lurks the Food and Drug Administration, to whose regulations it must also conform.

Despite these constraints, the industry has become a part of American life and has assumed its own place in the structure of production. It supplies those businesses into whose operations it has, over time, been integrated, and it will continue to do so as long as the population continues to buy those products so uniquely served by flexible packaging.

Its primary trade association is the Flexible Packaging Association. There are others, such as the Flexographic Technical Association, which has attracted some flexible packaging converters because of the importance of flexographic printing in their operations.

The industry, like other industries, has unique characteristics. Perhaps the most significant is the nature of what it converts. Plastic films, metal foils, and papers are the industry's bread and butter. These are essentially two-dimensional structures. The third dimension, thickness, may be as little as one thirty-thousandth of the width and is minuscule compared to the length, which is all but infinite.

The technology involved is concerned with the unwinding and winding of rolls of substrate through machines where inks, adhesives, or coatings are applied in patterns or over the entire surface. Tension on the web must be controlled, pressure and duration of cylinder contact (printing or coating, engraved or smooth) must also fall within defined limits and be held there with some exactitude, and total indicated runout of a pair of rolls should be of the order of 0.002″ or less.

The science is the chemistry and physics of surfaces, an area which only recently has become somewhat better understood through some of the newer methods developed for the study of surfaces (something stimulated by the needs of the electronic chip industry). Many of the materials used by the industry are polymers, which have a science of their own.

The art is that of *formulation:* the blending of polymers, slip additives, wetting agents, pigments, plasticizers, and other resins to achieve a desired result, whether it be film, ink, adhesive, heat-seal coating, or some other material.

Much of this process is left to suppliers, who do the basic work on inks, coatings, adhesives, films, grades of paper, etc. There are very few American converters who do any of their own formulating or build any of their own machines, in contradistinction to other parts of the world, such as Europe, where converting was begun by the film producers, due to quite different economic circumstances. Many European converters make their own adhesives, inks, and coatings; some, their own film; and a few design and build their own equipment.

The net effect is that in the U.S., most of the technological innovations adopted by converters are made by the machine makers, the ink makers, the adhesive and coating makers, the film makers, the polymer makers, and even some of the instrument makers.

BEGINNINGS

Toward the end of the 19th century and the beginning of the 20th, in the ongoing search for an "artificial silk," some major developments occurred, which were to have a profound effect on the existence and evolution of flexible packaging.

It was during this period that Cross, Bevan, and Beadle developed regenerated cellulose, and Camille and Henri Dreyfus developed cellulose acetate (despite the saying among chemists that he who "acetates" is lost). Both were used as textile fibers, sold under the tradename of *Rayon.* The manufacture and marketing of rayon became a worldwide industry. Later, the cellulose acetate fiber was sold simply as *Acetate* to distinguish it from regenerated cellulose, a different product.

In 1911, Jacques Brandenberger in France obtained a patent covering a continuous roll of transparent film made from regenerated cellulose. A company was formed to manufacture and market it. The company was called La Cellophane and the transparent film was called *Cellophane,* adapted from the Greek, meaning cellulose glass.

At almost the same time, J. P. Bemberg in Germany, who had been making rayon by the cupro-ammonium rather than the viscose process, introduced its version of a regenerated cellulose film.

A disgruntled engineer from La Cellophane went to Belgium and persuaded the Belgian Sidac Corp., a rayon producer, to join in making a regenerated cellulose film.

Thus was the flexible packaging industry born.

It was very nearly a stillbirth.

With the arrival of World War I, progress in finding uses for this amazing new material was all but halted. While there had been some desultory wrapping of candies and other products in tin foil or embossed paper, flexible packaging had not really demonstrated itself as a mode of carrying a product to market. Although cellulose acetate was first made as a film as early as 1909, its use was devoted solely to the manufacture of photographic film base, replacing the cellulose nitrate that had been used previously for this purpose because of the acetate's superior fire-resistance. No thought was given, at that time, to its possible use as a wrapping material.

Flexible packaging was on hold, and remained there until the war came to an end.

POST-WORLD WAR I

With the return of a civilian economy, a number of U.S. companies began searching for new products, new markets, perhaps whole new businesses into which to expand. The boom of war was gone, and while operations had been profitable, the immediate post-war years were bad with respect to economic activity, and new ideas required exploration.

One company seeking new investment areas was DuPont. Already in the rayon and acetate business, DuPont sought to manufacture a regenerated cellulose film similar to that made by La Cellophane.

After attempting to do this for a few years, with insufficiently consistent results, DuPont obtained a license from La Cellophane to make and sell *Cellophane* (the trade name was included in the license). The DuPont

Cellophane Company was formed in 1924. A plant was built, a research organization was formed, a sales force created, and the project was launched with all the gusto and spirit of an entrepreneurial enterprise.

Prior to this, however, La Cellophane had begun promoting the sale of its film in the U.S. In this effort, it had appointed a number of distributors. These distributors were inherited by DuPont, but the inheritance was nonbinding. DuPont reappointed all the La Cellophane distributors with a single exception. For reasons lost to history the firm of Bern and Wachenheim, located in New York City, was not rehired.

Adolph Bern, head of Bern and Wachenheim, was not at all pleased by DuPont's failure to select his firm as a storer and seller of *Cellophane*. He hied himself to Belgium and obtained a license from the Belgian Sidac Company.

In 1927, DuPont purchased the Grasselli Chemical Company, a manufacturer of heavy, inorganic chemicals (sulfuric acid and the like). The president of Grasselli at that time was a gentleman named Ralph Warburg. Why DuPont let Mr. Warburg go is now unknown, but Mr. Bern recognized his skills and hired Mr. Warburg to run the "Sylvania Manufacturing Company," as his fledgling producer of regenerated cellulose film was called. With capital and energy supplied by Mr. Bern and skillful management supplied by Mr. Warburg, the possibility that Sylvania might become successful seemed greater than zero.

Cellophane, as made at that time, had a serious flaw. It was extremely water-sensitive. As the ambient humidity changed, its water content changed, increasing the size of a wrapper with increasing water content and shrinking it with water loss, leading to serious dimensional instability. Ideal water content was about 7%; much more led to sloppiness and stretch in the wrapper; much less, to brittleness and frangibility of the film. The film's characteristics were somewhat dependent on how much more—how much less. Its water-vapor permeability was high, making it useless as a protective wrapper.

DuPont, conscious of these difficulties, assigned some of its research scientists to work out solutions for them.

In 1927, William Charch and Karl Prindle obtained a patent with a very large number of claims (72) covering a moisture-proof cellophane, made by coating the film with a cellulose nitrate-based coating containing wax, which spewed to the coating surfaces after drying, producing a low water-vapor permeability and a more dimensionally stable film. Closure of a wrapper was done by *solvent-sealing*—wetting the coating with a dab of solvent in the area to be sealed and holding the wetted sections together until the solvent-softened coating "set" by solvent evaporation. The process was slow, cumbersome, and uncertain, but it served its purpose.

In light of what is known today about polymers, resins, and waxes, the work seems routine and hardly worth the large number of claims granted by the patent office. But, considering the knowledge and experience of the time, the work demonstrated a high level of thought and originality.

This new development left Sylvania in a position of being noncompetitive, so their research people developed a modification of the DuPont coating and obtained a patent on a heat-sealing coating by incorporating a thermoplastic, cellulose nitrate-compatible resin in the coating.

After considerable discussion, DuPont licensed Sylvania to make and market moisture-proof cellophane and Sylvania licensed DuPont to make and market heat-sealing film.

By 1930, there were two producers of regenerated cellulose film quite prepared to satisfy any customer's requirements. Flexible packaging had become a reality. The third ingredient in the trio of paper, metal foil, and transparent film had arrived.

Those printers and others who had handled the film made by La Cellophane eagerly seized the coated product. Letterpress inks, which were used for printing the film, adhered better. These inks, the improved dimensional stability, and the process of making closures by heat-sealing all increased its potential uses greatly.

Then came the Great Depression.

INITIAL GROWTH

The uses of cellophane, contrary to expectations (because of the serious economic slump), grew during this period and so did the use of flexible packaging. Transparency and protection from ambient conditions were potent characteristics that could increase sales despite severe economic circumstances. Many printers who had become accustomed to converting the film became cellophane specialists, confining most of their activities to packaging applications.

The packaging-machinery makers, seeing the possibilities of new markets, cooperated fully.

Baked goods of many kinds, cookie and cracker boxes, candies of all varieties, and a host of other products found their way into cellophane wrappers. A signal triumph for film wrapping came in 1933, during the very depths of the Depression, when the cigarette companies began overwrapping their packs for two reasons: first, it enhanced the appearance of the pack, and second, it increased the shelf life of the cigarettes. While this did not mean additional business for the converter, it added to the glamour of transparent film as a package and behaved as an ever-present advertisement for film use.

Some major companies began to look upon possible uses of a variety of polymers as packaging films. The Goodyear Tire and Rubber Company produced a film of rubber hydrochloride. It was sold under the trade name of *Pliofilm,* and was the first thermoplastic film that was heat-sealable. While cellulose acetate was also thermoplastic and had existed as a film for some time, it wasn't truly heat-sealable.

Pliofilm was also water-resistant with a relatively low water-vapor permeability without the use of coatings. However, it was hazy, had a greenish color and a characteristic odor which, while it did not, in most cases, migrate to the product packaged, was offensive to many users. Despite these disadvantages, it found an immediate and major use in the packaging of physicians samples of a multitude of pharmaceutical products.

Union Carbide attempted to make a film from hydroxyethylcellulose, but never brought it to market. Sylvania Manufacturing did bring out such a film and sold it as a wrapper for silverware, a product for which cellophane was not suitable because of the very small quantity of sulfur retained by the film from the viscose manufacturing process. While Union Carbide did not succeed with a hydroxyethylcellulose film, it was successful in manufacturing a paper on which a coating of polyvinylchloride (PVC) was applied by a calendering process. This was marketed as a bottle cap liner and penetrated flexible packaging only slightly. Both Celanese and DuPont promoted their cellulose acetate films in flexible packaging, but substantial uses were difficult to find and cellophane, and to a lesser extent Pliofilm, dominated the marketplace.

INNOVATION UPON INNOVATION

During the 1930s, some new food products were introduced, along with new methods of processing. Frozen foods, as developed by Clarence Birdseye, exemplified these new processes. Dehydrated soups as originally marketed by Thomas J. Lipton Company represented new products.

Lipton's dehydrated chicken soup was packaged in a heat-sealed pouch comprised of printed aluminum foil, paper, and an interior layer of Pliofilm providing an hermetic seal and a low water-vapor permeability.

On the food side, new mixing equipment, drying ovens (for the noodles), new filling equipment, and heat-sealing equipment for closing the filled pouch were made. On the package-making side it required the design and construction of pouch-making machinery which operated at a then respectable speed of 60 pouches per minute. The machines folded the web, heat-sealed and cut off the pouch in register with the print. The construction of all this was a tall order at that time.

The product and the package (produced by Reynolds Metals Company) were quite successful. The soup was sold in two envelopes contained in an attractive small cardboard box.

Another major innovation related to processed cheeses, which had a checkered history. Alan Abrams of the Marathon Corporation (later part of American Can Co.) patented cellophane coated with a polyisobutylene reinforced wax. The coated film was formed into an envelope by folding the web, carefully heat-sealing the wax layer and cutting the envelope in register through the seal so that the rear seal of one envelope and the front seal of the next were part of the same original seal. This had also been done with the Lipton soup pouch.

The envelope was opened and inserted into the wooden box which carried the product to its final destination. The hot cheese was poured into the open envelope. The heat from the cheese melted the wax coating and the air between the cheese and the molten coating would be driven out. When the coating froze, it tended to cling to the cheese and prevent the growth of mold, by not permitting oxygen to make contact with the cheese.

This method of packaging became the standard for the 5-lb. block from which smaller quantities were cut for sale. Package and product suited each other and the method proved economical despite the investment in new equipment needed by both the package and cheese makers.

Many meat products were wrapped in cellophane or Pliofilm. Some fresh produce was wrapped in cellulose acetate (fresh fruits and vegetables must be permitted to breathe, otherwise they tend to rot).

Meanwhile, other developments coming out of the chemical industry in the 1930s were destined to have long-term effects on many industries, including flexible packaging.

THE DEVELOPING SCIENCE OF SYNTHETIC POLYMERS

In 1933, Imperial Chemical Industries (ICI), the British chemical giant, funded E. W. Fawcett to investigate chemical reactions at very high pressures. Ultra-high pressures had become achievable with the development of pumps capable of reaching 45,000 psi by A. Michels at the University of Amsterdam.

An ethylene polymer had been made in Fawcett's laboratory as early as 1933 by the high-pressure reaction of ethylene and benzaldehyde. This had established the fact that polymerization of ethylene was possible.

Fawcett and his co-workers were seeking a less costly process, and by 1935 had found a method of producing a useful ethylene polymer at high pressures by adding a small trace of oxygen to the pressurized gas. A patent application was filed and British patent 471,590 was issued to Fawcett and co-workers on September 6, 1937. By this time, ICI had a pilot plant running, producing roughly one ton per day of a respectable grade of low-density polyethylene (PE). In September of 1939, a full-scale production unit was built by ICI, making PE a high-pressure process. All of the plant's output at that time was used for military purposes, World War II having begun in Europe.

In 1929, DuPont took on a young chemistry instructor from Harvard University named Wallace Carothers. The company hired him to establish a group within it doing fundamental research in polymers and polymerization. Appropriations were made, an area of laboratory space was set aside for Carothers and his group, and DuPont was in the business of developing new science.

But the economy changed drastically with the onset of the Depression. Attitudes changed and soon Carothers found himself working on assigned projects with a clear commercial end in view.

His first task was to investigate what might be done with divinyl acetylene, which the dyestuffs department found in one of its processes. One of the members of Carothers' group found chloroprene present as an impurity and noted that it polymerized spontaneously. Carothers and his people learned to synthesize it and control its tendency toward spontaneous polymerization. Plants were built for the manufacture of both chloroprene and its polymer, the first commercial synthetic rubber marketed in the U.S., sold to this day under the trade name *Neoprene.*

Carothers was then assigned the project of making a truly synthetic fiber, silk-like, if possible. Both rayon and acetate were not considered satisfactory; they were much too water-sensitive to be washable except with care. But they were inexpensive, easily woven, dyed, and sewn, and were adaptable to virtually every garment, from ladies' slips to men's suits.

But they were really relatively simple derivatives of cellulose fiber, a primary natural product, whose properties were as consistent as the vagaries of nature permitted.

Carothers turned his attention to *C* polymers, including those that were polymerized by condensation, such as polyesters formed by reacting a dibasic acid with a glycol, or a lactone with itself; or polyamides made by reacting a dibasic acid with a diamine or a lactam with itself.

The polyesters Carothers made were not suitable as fibers, but some of the polyamides were, and in late 1936 and early 1937 Carothers made a successful fiber from a polyamide polymer synthesized from hexamethylene diamine and adipic acid, both of which, at that time, were laboratory curiosities.

DuPont proceeded to put together a production unit to make *Nylon,* as the material was named. In addition, production units were built to manufacture hexamethylene diamine and adipic acid in sufficient quantity to support what was expected to become a major product in commerce. After an initial production of some nylon yarns for ladies' stockings, the entire output of polymer was reserved for military uses.

By 1941, the clouds of war were low and the lights were going out.

Carothers' work, like Fawcett's, was primary. It introduced, among other concepts, the notion of *superpolymers,* which were made by driving the product of a condensation reaction to very high molecular weights, strongly influencing much of the later work on engineering plastics. His classification of polymers into *A* types (those polymerizing by addition) and *C* types (those polymerizing by condensation) still stands today.

Wallace Hume Carothers, one of the truly great chemists of his or any time, did not live to see the success of his work. He committed suicide in 1937.

In the mid-1930s, a patent was issued to Irving Gurwick for rotogravure printing of color on color on a transparent film, so that when the film was turned over, the printed legend would read correctly. Thus, the printing was protected from scuffs, spills, and other environmental insults by the film itself. This was a clever way of taking advantage of the transparency of cellophane, the film most used at the time, and the one which lent itself readily to printing because of its stiffness and heat resistance. This process was called *reverse printing.*

Some of the major converters of the time recognized the significance of this development. One of them, the Shellmar Products Company, bought Mr. Gurwick's company, thus gaining possession of Mr. Gurwick's patent and Mr. Gurwick.

Shellmar Products licensed some of the other major converters, such as the Dobeckmun Company of Cleveland and Milprint, Inc., of Milwaukee. The club of licensees was able, to a limited extent, to hold the line on price, and increase their profit margins. They were also able to stop infringers through their cooperation. For a time, some order was maintained within the industry.

Then came World War II, and once again, flexible packaging was put on hold.

FLEXIBLE PACKAGING AT WAR

Complex laminations were developed to suit the needs of armies. Structures made of fabric combined with paper, asphalt laminated to aluminum foil which was then coated on the foil side with a mil of a polyvinylbutyral

(PVB) formulation to give it additional web strength and very strong heat seals were developed and used for the packaging of spare parts for planes, tanks, and artillery.

A flexible material using kraft paper, asphalt laminated to aluminum foil, and a mil of PVB (which, until then, had been used as a safety-glass plastic) coating for strength and heat-sealability was made. This was used, in the form of envelopes, to wrap boxes of 50-caliber machine-gun bullets.

With cellophane replacing the PVB, the structure became a ration package.

These laminations were rugged enough for the packaged items to be thrown overboard during a landing operation and salvaged with the product packaged, whether ration or armament, dry, in its original form, and usable.

Sylvania Manufacturing produced a polyvinylidenechloride (PVdC, Saran) coated cellophane for the wrapping of the cork plugs used as ballast for loading artillery shells. Of all the products made for military packaging, this was the only one, which, after a brief hiatus, entered the civilian economy after the war's end.

ANOTHER WAR ENDS

With the end of hostilities and the restoration of a normal society, civilian need for everything from homes and cars to meats and cheeses was overwhelming and the problems for all industry became those of producing enough to satisfy this demand.

ICI had an arrangement with DuPont, whereby DuPont was licensed to produce and market ICI patented products in the U.S. and ICI was licensed to produce and market DuPont patented products outside the U.S. In some areas of the world the two competed, largely in minor markets.

Thus the two chemical behemoths exchanged nylon for PE. DuPont, in turn, sub-licensed Union Carbide to manufacture and market ethylene polymers. During the very late 1940s and early 1950s, the price of PE steadily dropped. A host of film extruders sprang up, including the Visqueen Co., whose parent, the Visking Company, had been a major customer of Union Carbide as a purchaser of *Visten* film for sausage casings. Visqueen soon became a major customer of Union Carbide for PE polymer (flake, as the pellets were called) as the demand for PE film continuously grew.

Starting in the early 1950s, the use of PE began a period of unprecedented growth still going on today. And with it flexible packaging entered an era of frantic expansion. The PE bag, printed and unprinted, was used for everything from screws to stockings, from clamps to shirts. Bags were used by some retailers to hold whatever the retailer thought might look better in a bag.

One of the results of this rapid growth was that getting into converting required a relatively low investment. All one needed was a modicum of courage, a PE bag machine, and some ability to sell (or a partner who could) and one became a converter. The industry burgeoned with newcomers, many of whom remained in it and survived profitably.

A second piece of equipment frequently purchased by these entrepreneurs was a printing press; since a flexographic press was adequate and rotogravure presses were expensive, flexography became the printing process of choice for the industry and has remained so.

In his search for fiber and film forming polymers, Carothers had synthesized many polyester polymers, but had neglected the polymer formed by the condensation of ethylene glycol and terephthalic acid. ICI acquired a patent on such a polymer and, of course, DuPont obtained a license to manufacture the material in the U.S.

Initially a fiber was produced and sold under the name *Dacron.* As a fiber, the polymer was quite successful in blends with cotton for shirts and blouses and with wool for suits. But it also formed a film and if this film were oriented biaxially (i.e., stretched laterally and longitudinally to align some of the polymer chains laterally and some of them longitudinally and then heat set to stabilize it), a film of remarkable properties resulted: strong, elastic, with a high modulus of elasticity making it relatively stiff compared to other plastic films, remarkable clarity, and moderately low transmission properties, both of water-vapor and gas. When this film was PVdC-coated, it provided quite good barrier properties. Its strength was such that a 0.0005″ thick film was stable enough and strong enough to be processed.

The film was sold by DuPont under the trade name *Mylar.* Today, there are two similar films: *Celanar,* produced by American Hoechst (obtained when Hoechst bought the Celanese Corporation), and *Melinex,* produced by ICI.

THE END OF EXCLUSIVITY AND THE ONSET OF MULTI-LICENSING

In 1954, a federal judge ruled that ICI must make licenses to manufacture PE available to all who met the requirements for such a license. Spencer Chemical (later Gulf Chemical, a wholly owned subsidiary of Gulf Oil),

U.S. Industrial Alcohol (later U.S.I. Chemicals, still later Quantum Chemical), Monsanto Chemical, Dow Chemical, and others obtained licenses. Sources of PE polymer and film became plentiful.

Also in 1954, Dr. Karl Ziegler, who had worked for some time on catalysts to produce a low-pressure polymer of ethylene, having obtained basic patents on these—at that time, strange aluminum alkyl, titanium trichloride, and tetrachloride complexes. He found that the polymers made by his low-pressure process, and using his coordination catalysts, as they came to be called, were quite different from the low-density products of commerce. Their density was higher. Films formed from them were more linear in nature, their chemical resistance was superior, their melting points were higher.

A similar set of products had been made by Phillips Petroleum using metal oxides as catalysts. Licenses were obtained by most of the PE producers from both Dr. Ziegler and Phillips. Ultimately, it became possible to control the density of the polymer by choice of catalyst, process variables, and batch blending, and it became possible to specify polymer density from 0.912–0.960; and molecular weights (as measured by an arbitrary dimension known as *melt index*—the lower the melt index, the higher the molecular weight), from a fractional value of melt index to one of double digits—from very high molecular weights to very low ones.

DuPont (and Sylvania) had a policy of giving a 7% discount on cellophane sold to converters. It had been used selectively, freezing some quite legitimate converters out of eligibility for the discount. In 1955, a U.S. government ruling was handed down making all converters eligible for the discount. Thus, 1954 and 1955 were important years for the converter, although many never realized it.

LAMINATIONS—NEW APPLICATIONS

It was in the late 1940s and early 1950s that laminations began to enter the marketplace in a major way. One of the earliest was a combination of cellulose acetate and Pliofilm for the packaging of pharmaceutical samples for medical detail-men to distribute to physicians. Such packages were made on an Ivers-Lee type machine, usually by contract packagers—commercial organizations whose business it was to package other people's products.

A favorite laminated structure of the pharmaceutical industry was one consisting of cellulose acetate/aluminum foil/Pliofilm because of its almost zero permeability to water vapor and noncondensable gases. A favorite food application was the packaging of single servings of soluble coffee.

An important new development at that time was the flexible vacuum package, which started out as an envelope made from cellophane (for the low gas transmission) and Pliofilm (for the hermetic seal). Initially used for smoked meats such as bacon, it soon expanded into gas-packing of sliced natural cheeses (flushed with nitrogen). It was found that the package extended the shelf life of the products sufficiently to enable the packer to ship them a considerable distance from his locality. Of course, the widespread use of refrigeration in the supermarkets, which were then in their infancy, made the venture possible.

With the dramatic reduction in price of PE film and flake, PE replaced Pliofilm as the sealing layer. Cellophane temporarily strengthened its grip on the gas barrier part of the lamination with the introduction of PVdC coatings replacing the cellulose nitrate-wax coatings that had been in use since 1929.

Along with the growth of supermarkets, the 1950s saw an increased use of flexible packaging as a way to distribute groceries and other packaged items. A bag in a box became a popular sales package; the box supplying rigidity and ease of stacking on shelves for display, and the bag providing protection from moisture, oxygen, and other invaders.

Concomitant with this came the introduction of the plastic film shopping bag, printed and unprinted, with and without handles, many with cutouts by which to grip them. In America, these have become ubiquitous.

Flexible packaging was looked upon with favor by the financial community and many major companies purchased converters. Continental Can bought the Shellmar Products Company. Dow Chemical bought the Dobeckmun Company. Philip Morris bought Milprint, Inc. American Can bought the Marathon Corp. Bemis Bag bought Curwood, Inc. W. R. Grace bought the Dewey and Almy Chemical Company. Union Carbide bought Visqueen, a major PE film producer, and then was forced to disgorge it by the Federal Trade Commission; apparently, one was able to buy one's supplier but not one's customer. The litany goes on.

Many converters became publicly owned: e.g., Cellucraft and Package Products. And there were and are others. The industry had matured.

EXTRUSION COATING

During the late 1940s and early 1950s, extrusion coating, a process whereby a molten film of polymer (e.g., PE) could be coated on a substrate and, very nearly, serve as a laminated film, was introduced to industry. It offered a lower cost than lamination, because the user of the process was able to start with pellets of polymer and form his own film in situ rather than purchasing a preformed film, and paying someone for forming it.

Machinery for the process was designed, built, and sold by the F. W. Egan Co., Black-Clawson, and others. Machines were purchased by some of the major paper companies and PE-coated paper became an integral part of virtually all multiwalled bags. Many converters also entered the field and a major new product was developed: a paper/PE/aluminum foil/PE structure, in which the two layers of PE were extruded. The structure was used for dehydrated soups and pet foods, among other things.

Adhesion of the extrudate to aluminum foil and paper was accomplished by extruding at a temperature high enough (about 600°F) to make incipient oxidation of the extruded PE possible by drawing the polymer down in a sufficiently long air gap (perhaps 6 inches) to permit such oxidation to occur. In the cases of films such as nylon, polyester, cellophane, et al., it was necessary to apply a thin film of an adhesion promoter (and dry it) to obtain satisfactory bonds.

Although bonding to aluminum foil and paper occurred if extrudate temperature were high enough, sometimes the requirements of a product were such as to necessitate use of an adhesion promoter for these substrates.

By the early 1960s, the standard product for the packaging of dehydrated soups was a combination of paper/PE/aluminum foil/PE, where the two layers of PE were extruded. Before the decade ended, some of the major producers of this composite placed two extruders in tandem and made the four-ply structure in a single pass.

STEREOSPECIFIC POLYMERS

In the mid- and late 1950s, Giulio Natta and his co-workers developed stereospecific polypropylenes at the laboratories of Montecatini, then a major Italian chemical company using Ziegler-type catalysts.

Polypropylene can be polymerized in a form where all the methyl groups lie on one side of the chain backbone. This is known as *isotactic* polymer. It can also be made in a form where the methyl groups alternate on which side of the backbone they can lie. This is known as *syndiotactic*. And, it can be made with a random location of the methyl groups. This is known as *atactic*. The form is determined by the choice of catalyst.

The initial publication of Natta's work occurred in Italy in 1954. In the U.S., an article by Natta and co-workers appeared in the *Journal of the American Chemical Society* in 1955.

It was Madame Natta who invented the names isotactic, syndiotactic, and atactic.

Polypropylene is the lightest of all polyolefins in good supply today, its density being 0.905. For every 100 lbs. of crude petroleum put through a Houdry catalytic cracker, one pound of propylene is formed as a by-product. Its low density and promise of low cost stimulated strenuous efforts by oil and chemical companies to solve many of the problems associated with its manufacture and use.

It was found that a mixture of about 65% isotactic and 35% atactic polymer processed well, i.e., it was readily extruded into film. The presence of a tertiary carbon atom made the film too readily oxidizable. The orientation of the film presented special problems in the need for very close control of temperatures of stretching and heat setting, and maintenance of uniform tension in the oven. All these problems took time to solve and it was not until the mid-1960s that the film really began to be used in any quantity.

The major players in the polypropylene game at that time were Hercules and Mobil Chemical, a subsidiary of Mobil Oil. They were later joined by Exxon, Crown-Zellerbach and Northern Petrochemical. Exxon and Crown-Zellerbach did not make oriented film, Northern Petrochemical did.

Mobil Chemical, which had gotten into the polyethylene bag business through the purchase of the Kordite Corporation, a major producer of these products, also acquired the polypropylene business of the Diamond-Shamrock Corporation. This was consolidated with the Kordite operation at Macedon, NY, and Mobil proceeded with intensive work on coatings with various polymers, including acrylics for heat-sealability and PVdC to add barrier properties to the film. Hercules proceeded in a similar direction. The markets polypropylene film was directed at were those that cellophane had dominated.

MACHINE PACKAGING OF SNACK FOODS

Oriented, PVdC-coated polypropylene has displaced cellophane all but completely. A primary impetus in polypropylene's rise as a major product for the converter was the development of the form-fill-seal machines for the packaging of potato chips and other snacks. Woodman, Mirapak, Stokes and Smith, Transwrap, and others made such machines and major snack food producers such as Frito-Lay, Wise, Laura Scudder, Dandee, and many others bought them.

It was necessary to devise a structure that would run on these machines and yield a satisfactory package. One of the first was a lamination of PVdC-coated cellophane with itself. One layer of cellophane was soon replaced by a layer of oriented polypropylene (OPP) with the cellophane as the interior layer to serve as a barrier to any retained solvent in the adhesive. This was soon replaced by two layers of PVdC-coated OPP laminated together. This has been the package of choice up to the present day.

It has been replaced by a clear OPP, printed with an overall white to render it opaque, laminated with PE to a metalized OPP. For corn chips and snacks other than potato chips, the package has become two layers of OPP laminated together with a PVdC adhesive. Bonds leave something to be desired, but cost to the packer is low and the package does not fall apart in normal handling.

CONSOLIDATION

By the end of the 1960s, the materials of converting were all present. Some specialty materials were missing, but those that are used in quantity today were all available to the converter.

The 1970s were devoted largely to machine improvement. The application of sophisticated electronic circuitry to converting machinery became the order of the day. Drives with automatic tension controls were devised by Reliance Electric Company, the Louis Allis Company, and others. Winding and slitting machines were built that could be programmed to wind a roll in accordance with a pre-set applied tension pattern. Automatic registration controls on rotogravure presses became commonplace. And many converters put operations in-line. It became usual for a converter to have one or more machines designed to print and laminate in a single operation. Tandem extruders have already been mentioned.

As the structures sold by the industry became more complex for more and more specialized applications, the trend toward having more operations performed by a single piece of equipment increased, not so much because this reduced the labor cost of the product (which it did), but, more importantly, because it reduced materials waste.

One process of this type which had been developed, though not used widely, in the late 1950s, was co-extrusion: the extrusion of a number of different polymers which are joined together into a composite film, perhaps one with a core of high-density polyethylene encased in two layers of low-density PE. An example of a commercial product that is not totally polyolefinic is Saranex, a thin layer of PVdC with a layer of PE on either side, marketed by the Dow Chemical Company.

Some of the machine makers have suggested the notion of co-extrusion coating and the F. W. Egan Company has designed and built such machinery. There are others, such as Black-Clawson.

In the early 1960s, a process was developed by Milprint, Inc., for obtaining substrate adhesion of an extrudate of PE at lower temperatures than were customarily used. Adhesion of extrudate to substrate was obtained without oxidation of the film in the air gap. The process also applied to other polyolefins which required such short drawdowns to obtain a good extrudate that there was too little time of exposure in the air gap to get incipient oxidation and the extrudate failed to adhere to the substrate.

Polypropylene is such a polymer, but by use of the "Nealam process" (as the method was called), an extrusion-coated polypropylene with adequate adhesion could be obtained.

TWO-PART CURING ADHESIVES

During the early 1950s, Tee-Pak, a producer of sausage casings, developed in its laboratory a two-part product consisting of an isocyanate and a polyol which, when reacted, yielded a polyurethane with remarkable adhesion to virtually all substrates.

Not being in the converting business, Tee-Pak lost control of its finding. DuPont, through its Fabrics and Finishes Department, marketed a urethane prepolymer which, when reacted with an amine, formed a polyurea that yielded strong bonds when used to laminate flexible packaging webs. The urethane prepolymers (later sold to Uniroyal) were sold under the name *Adiprene* and a particular amine under the name *Moca.*

DuPont also developed a line of terephthalic acid polymers of low molecular weight and with an excess of hydroxyl groups. These polymers reacted readily with isocyanates or isocyanate prepolymers, forming a strong bond between two substrates, e.g., polyester and PE.

Adcote, Polymer Industries, National Starch and Chemical, and others began marketing the two-part isocyanate-polyol or isocyanate-amine adhesives to the converters. Initially, the response was negative, but over time, they came to be accepted because of the superior bonds obtained. To reduce costs, one-part moisture curing adhesives were marketed, but these never gave the bonds, chemical resistance, or heat resistance of a two-part polyester-isocyanate adhesive. Adhesives which reacted in situ to form polyurethanes and/or polyureas became the standard in the industry for almost all laminations.

SURFACE TREATMENT OF SUBSTRATES

PE and other polyolefins are so inert chemically that in order to get bonding to the surface it is necessary to treat it.

Initially, the treatment used was a chemical oxidation of the surface. This was slow and hazardous. In 1953, Kritchever introduced the idea of flame-treatment to oxidize the surface of the film and thus produce adhesion. Traver obtained a British patent covering a process of treating a film by a corona discharge which has become the standard of the industry.

Surface treatment, a revolutionary idea then and an obvious one now, represents one of those rare occasions when the burden of obtaining a satisfactory result is not solely the responsibility of the formulator, but is shared with him by a process for producing active sites on the surface to which bonding takes place.

Today, the use of corona discharge or plasma treatment has become so widespread that many converters re-treat polyolefin films and treat almost all the substrates they process, even aluminum foil.

FURTHER DEVELOPMENT IN VACUUM PACKAGING

Vacuum packaging of meats and gas-flushing of natural cheese is now accepted practice, the original pouch having been replaced by a two-web, automatic operation. One web which is formed to accommodate the thickness of the product is made from nylon/PE. The non-forming one is made from polyester/PE. For very high barrier requirements, the nylon and polyester films are PVdC-coated.

TODAY—THE CONVERTER AND THE ENVIRONMENT

The attention of the converter today is drawn to the problems involved in conforming to the regulations of the various federal, state, and local environmental agencies. The converter seems to be considered a primary polluter; vapors from the solvents for inks and from other fluid coating materials pollute the atmosphere. Underground solvent storage tanks develop leaks, permitting the solvent to seep into the ground, poisoning the soil and ultimately getting into the groundwater, rendering it unfit for human consumption. And the films used are perhaps the worst. None of them, with the exception of cellophane, are biodegradable, and the coatings used on cellophane are not.

The case is very much overstated. The converter is not a villain. The converting industry is no worse a polluter than any other industrial operation, and is making an honest effort to meet all the environmental requirements, all the FDA regulations, all the Federal Trade Commission requirements and the state, county, city, village, or borough rules, while, at the same time, continuing to supply customers with the quality of product they have learned to expect.

It's a tall order.

It is unfortunate that there are so many rules, but perhaps they are needed. In our history, regulations have usually arisen because they were necessary. But they always seem to go too far.

In the case of the converting industry, progress is being made. Storage tanks are being built above ground with wells around them to catch leakage. Solid waste is being transported to toxic waste sites rather than ordinary landfills, and, so far as solvent emissions are concerned, there are waterborne inks and adhesives available. Some converters prefer incineration equipment, others prefer the 100% solids route. There are alternatives.

It will be costly, initially.

Meanwhile, there are new developments. The use of metallized films with both polyester for coffee packages and polypropylene for snack foods is increasing. Laminations of polyester/cast polypropylene are entering the hospital supply market as half of a gas-sterilizable package and all of a steam sterilizable one. A combination of polyester/aluminum foil/cast polypropylene is making inroads in whatever market there is for retort pouches (the structure withstands the rigors of steam sterilization). Hospitals tend to welcome what the converter offers. The consumer has not been overwhelmed by the quality of steam-sterilized foods.

Developments not mentioned in their proper time include the boil-in-bag for vacuum-packed frozen vegetables, soups, and other products. This is a way to preserve original taste and texture–packaged in polyester film combined with medium density PE (0.935 or thereabout) for better oil resistance and a higher melting point than low-density PE. In the development of Scotchpack by 3M, a combination of polyester and PE, the bond between the polyester and the PE is made by exposing the bonding surfaces to irradiation by UV prior to combining the two. In the laminated or extrusion-coated products made by the converter, a two-part polyol-isocyanate adhesive or primer is used for improved bond strength and chemical resistance. In the 3M product, there is no need for this.

No mention was made of shrink wrapping with a variety of oriented, non-heat-set films because this development did not involve the converter.

When an industry is doing $300 million of business annually, as this industry did in the mid-1950s, and a new $30 million annual market is found, it is hailed as a major breakthrough (a much abused word), and perhaps it should be. After all, such a market is 10% of the total volume.

But when the industry is doing $3 billion annually, as flexible packaging is doing today, such a new market is interesting but no breakthrough; it is now only 1% of the industry total.

In any case, much of the glamour and excitement of the early decades, when every innovation was significant and there were so many, so very

TABLE 1.1. Industry Sales in Recent Years (SIC 2671).

	In Millions of Dollars			
	1987	1988	1989	1990 (est.)
Value of shipments	2,433	2,633	2,619	2,887
Value of shipments (1987 dollars)	2,433	2,494	2,596	2,672

many, has vanished. This happens to all industries as they grow older, requiring those who have chosen participation in an industry as their way of life to be conscious of those changes demanding product innovation and imaginative marketing.

In 1987, the U.S. Department of Commerce, in order to present a more orderly pattern of statistics, created SIC (Standard Industrial Classification) 2671, which covers printed, coated, or laminated paper, film, and foils, used only for packaging. The resulting set of numbers is as authoritative as any available for the flexible packaging industry as a totality of its economic activity. Some of the recent figures are shown in Table 1.1.

SECTION ONE

THE UNIT OPERATIONS OF CONVERTING

CONVERTING of films, papers, and metal foils may be broken down into individual operations such as printing or coating, which stand by themselves as a unit. In each job one or more of these operations is performed. Not everything the converter makes requires all of them, but they comprise the "unit operations" of converting, to borrow a phrase from chemical engineering.

These unit operations are:

- Unit Operation 1—Surface Treatment
- Unit Operation 2—Printing
- Unit Operation 3—Coating
- Unit Operation 4—The Drying of a Moving Web
- Unit Operation 5—Laminating
- Unit Operation 6—Roll Winding and Slitting
- Unit Operation 7—Heat Sealing

This breakdown of the technology of converting assists in conceiving of converting as the sum of the processes used to serve the needs of the product and its packager.

A converting machine consisting of a printing press in-line with a coater-laminator (or, alternatively, with an extrusion coater) will produce a finished roll, which can be slit to a given size and sent to a pouch-maker. Starting at the beginning, with surface treatment of the substrate on the press and going step-wise through the machine the slitter and the pouch-maker, this roll will go through all the unit operations of converting.

CHAPTER 2

Unit Operation 1 – Surface Treatment of Substrates

WITH the advent of polyethylene (PE) into the marketplace following World War II, many changes occurred in both the plastics and the converting industries. Never before had there been a material with so many useful properties and with so strong a possibility of very low cost.

The first film applications were simply bags in which almost anything could be packaged, from men's shirts to hardware, screws to handkerchiefs, wall brackets to screwdrivers. Anything that might look or sell better was packaged using PE.

ADHESION TO PE

But soon, problems arose in the printing of this extremely promising material. Because of its inertness, nothing, particularly inks, in which all converters were vitally interested, adhered to it. And despite heroic efforts on the part of the ink makers, no solution to the problem seemed at hand.

The material had a very low critical surface tension: 31 dynes/cm. Fluid inks, such as those used by the converter for flexographic or rotogravure work generally had, and still have, surface tensions in the 34 – 38 dynes per cm range. Some of the water-borne inks in use today have surface tensions even higher.

Instead of shrugging their shoulders and accepting defeat, many individuals from both the film and converting industries responded with the thought that a solution could be found in treating the film in such a way as to make the surface wettable and adherable, rather than depending on the ink formulator's art.

Initially, efforts were made to do this chemically by oxidizing the surface using agents like chromic acid in solution or exposure to high concentrations of chlorine or ozone, with or without simultaneous exposure to ultraviolet radiation. There was little doubt that an oxidized PE surface wet readily and showed good ink adhesion but the processes used to get the desired result were cumbersome, dangerous, and costly.

After a few years, M. F. Kritchever found that by passing the film through a gas flame, enough surface oxidation occurred to produce wetting by and adhesion of inks (U.S. patent 2,648,097, August 11, 1953). Soon, an apparatus became available that would burn the PE just enough to allow ink adhesion.

But, it was not totally satisfactory. Controlling the flame required constant attention, and maintaining treatment level (as measured by surface tension values) was difficult. G. W. Traver obtained British patent 771,234, dated March 27, 1957, by describing a process for activating the surface of PE by passing the film through an electric field where it was bombarded by ions and electrons – in other words, a plasma or a corona discharge (CD). This process of film treatment has persisted to the present and is likely to continue, for while the equipment to treat film by this method has changed mechanically since its introduction, the electrical principles have remained the same.

CURRENT CD OR "PLASMA" TREATMENT

Today, CD treatment of film has become standard among the makers of polyolefin films and many converters. It has extended its application to almost all materials used by converters, and should become a universal practice among converters to treat whatever material may be in process prior to printing or coating. Since treating stations are available as complete, separate units, they are easily installed for in-line operation.

While CD treatment can be done using a 60 Hz electrical source (and it frequently is in laboratory work), by using a higher frequency, the impedance of a largely capacitive electrical circuit is reduced, and a treater circuit is totally capacitive until the corona begins. The power factor of the circuit is higher and more of the current goes into useful watts. For this reason, frequencies of 3,000 – 30,000 Hz are in use for the bulk of CD equipment available, and power supplies are sold jointly with the treating station.

Initially, higher frequencies than 60 Hz were produced by spark-gaps or multipolar generators. But very quickly vacuum tube and then solid state devices replaced the earlier machines. Today, all equipment generating the frequencies needed uses solid-state circuitry. Some include computerized controls for impedance matching with the load. The power supply includes a high voltage transformer to yield the thousands of volts needed to initiate a corona. Flame treatment is still used for molded articles requiring decoration where sections are so thick that an inordinately high voltage would be needed to start a corona, making CD treatment impractical.

DESCRIPTION OF APPARATUS

Figure 2.1 shows, schematically, a typical CD unit which might be attached to a film extruder in a film production plant or a converting machine such as a printing press or coater-laminator in a converting plant.

The high voltage side of the power supply is connected to a metal electrode placed over a metal roll covered with an insulating layer. Any of several insulators may be used, including silicones and Hypalon. These are placed with an air-gap between the electrode and the insulation-covered metal roll. The substrate to be treated is run over the insulation-covered roll and then passed to the next production stage – printing, coating, or laminating. Treatment is performed in the air gap. When the voltage is high enough to start a corona, the air ionizes, becomes conductive and carries enough power to treat the substrate.

Today, the electrode might be one of the newer ceramic electrodes, where the insulation is moved from the metal roll carrying the substrate to the electrode. The ceramic material might be a true ceramic, quartz, glass, or a ceramic-type material dispersed in an elastomer of plastic material. Using a ceramic electrode eliminates the need for the insulation cover on the metal substrate carrier, which therefore can be left bare. In any case, the elements of the electric circuit are: ground, transformer, electrode, air gap, substrate, insulation, metal roll, and ground.

Today, units are delivered with a transformer to supply the high voltage incorporated in the power supply. Figure 2.2 shows the electrical circuit matching the mechanics of Figure 2.1. The transformer is not shown as separate in part (b), the circuit for a modern ceramic electrode.

The insulation may be between the substrate and the metal roll or it may be between the electrode and the air gap as in the case of ceramic electrodes. The insulation is needed to protect the apparatus once a corona begins and the voltage in the air gap drops to the steady state voltage of 500 – 600 volts. This voltage varies with humidity of the air; the higher the humidity, the lower the voltage needed to sustain a corona.

When using a ceramic electrode, the circuit becomes: ground, high-voltage power supply, electrode, ceramic insulation, air gap, substrate, bare metal roll, ground.

Ceramic electrodes are made in a number of ways. The simplest is to fabricate a tube of the ceramic desired and then to insert a tube of aluminum

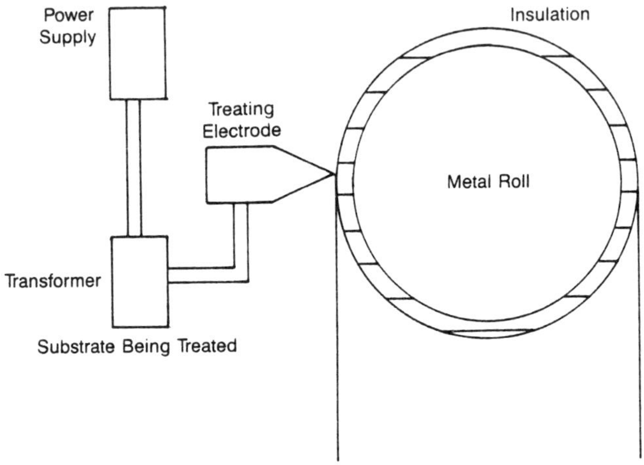

(a) Insulated metal roll—transformer

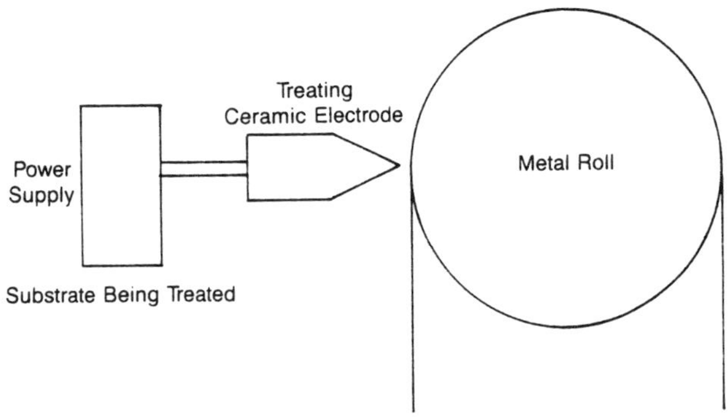

(b) Ceramic electrode—high voltage power supply

FIGURE 2.1. Mechanical arrangement of treater station schematic: (a) insulated metal roll—transformers; (b) ceramic electrode—high voltage power supply.

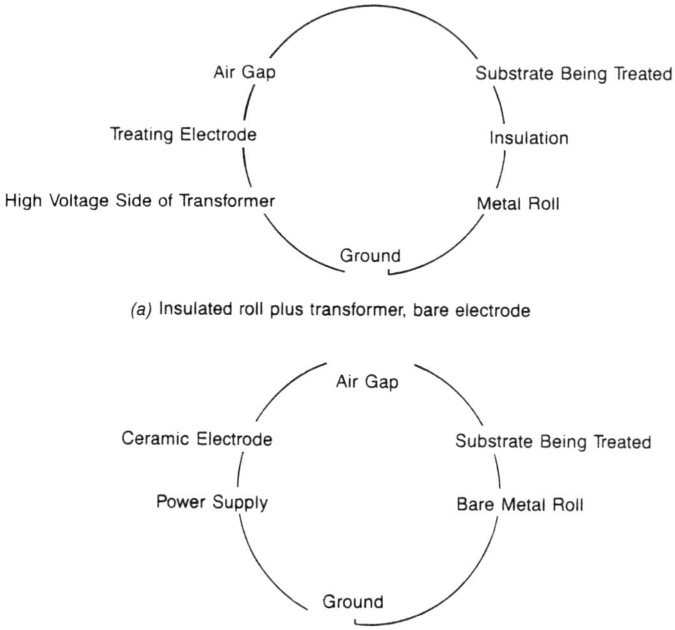

(a) Insulated roll plus transformer, bare electrode

(b) Bare metal roll, ceramic covered electrode, high voltage power supply

FIGURE 2.2. Electrical circuit accompanying Figure 2.1: (a) insulated roll plus transformer, bare electrode; (b) bare metal roll, ceramic covered electrode, high voltage power supply.

to act as the conductor. This has the disadvantage of not always obtaining a snug fit of the aluminum inside the ceramic; it is difficult not to leave air spaces.

A frequently used method is to make whatever shape may be needed out of the ceramic material, deposit a thin layer of metal (vacuum metallizing or other method), sprinkle in a thin layer of aluminum powder, and then place an aluminum plate over the powder. A spring is welded to the plate, and the apparatus built so that the spring is constantly forcing the plate against the powder, making a good electrical contact. The ceramic material is the outside of the electrode. Such units may be long or short, mounted in a track and used in such a way as to adjust the width of the electrode to conform to the substrate width. This construction of a treater using such a device is described in U.S. patent 4,533,523, granted to Andreas Ahlbrandt on August 6, 1985.

Another method involves using a dispersion of an inorganic material such as aluminum oxide, olivine, quartz, some other form of silicon dioxide, or a silicate dispersed in a silicone. This is molded around and cured to an aluminum tube. This is not totally ceramic but it has shown itself to be functionally almost equivalent to a truly ceramic device.

ELECTRICAL CIRCUIT

Figure 2.2 shows the electrical circuit equivalent to the mechanical scheme illustrated in Figure 2.1 (it does not show the power supply connected to the low-voltage side of the transformer). As already noted, the high-voltage side of the transformer is connected to the electrode.

On the start-up, the circuit is all but completely capacitive. The only current flowing is that due to stray ions and electrons probably produced by stray radiation and thermal collision of gas molecules in the air gap. However, this is so slight as to be barely detectable even with the most sensitive instruments. As the voltage rises and ionization of the air gap increases, a point is reached where a hissing noise may be heard, a blue-violet glow appears in the air gap, a steady flow of current occurs and the corona has begun.

The field strength required to produce a steady CD in the air gap is 70,000 volts per inch. This varies with humidity and temperature but is an acceptable value with which to perform some simple calculations. Hence, to start a corona, it will be necessary to apply $70{,}000 \cdot h$ volts across the gap between electrode surface and substrate being treated where h is the width of this gap in inches.

If the air gap is, let us say, 1/8 inch (125 mils), the voltage needed to initiate a corona is 8,750. Once current flows through the gap, the voltage drops to 500–600 volts, leaving the balance of the voltage as the difference in potential between the surface of the substrate through the substrate, the layer of insulation, the metal roll and the ground – that is, the potential difference between substrate surface being treated and the ground.

If the potential across the substrate exceeds the breakdown voltage of the substrate, sparks will fly, burning holes in it, perhaps damaging the insulation and causing a good deal of the current to short circuit, a totally undesirable result. It is vital that the air gap be kept small (not over 150 mils) and that the thickness of the layer of insulation on the metal roll (or the ceramic covering a ceramic electrode) be sufficient to prevent breakdown of the circuit.

The air gap, all substrates a converter is likely to encounter (with the exception of aluminum foil, conductive plastics, and some conductive inks), the ceramic on a ceramic electrode, and the insulation on an insulated roll are all dielectrics. In any dielectric subjected to an electromagnetic field, the electrons of the atoms comprising the dielectric are pulled in the direction of positive applied voltage, changing direction with each alternation of the current. This is known as electric displacement, or the displacement current, a concept used by James Clerk Maxwell in developing his famous electromagnetic field equations.

To make a calculation of the voltage drops across the elements of the electric circuit formed by electrode – air gap – substrate – insulation – ground – high-voltage side of the transformer – electrode, it is necessary to know the dielectric constant of each element of the circuit and its thickness. The voltage drop across each is inversely proportional to its dielectric constant and directly proportional to its thickness.

It is only the dielectrics in the circuit to which this applies: air gap, substrate, insulation. (The insulation might be the ceramic coating on a ceramic electrode.) This does not alter the method of calculation.

Using 70,000 volts/inch as the breakdown voltage for air, a 1/8 inch air gap (125 mils), and a dielectric constant of 1 (for air), a 2 mil substrate (PE) with a dielectric constant of 2, 1/8 inch (125 mils) layer of insulation with a dielectric constant of 5, we have all the necessary facts to find what voltage will be needed to produce a corona, and if the circuit is a safe one.

For a 125 mil air gap, a voltage of 8,750 will cause the air to ionize enough to become a conductor. Using the numbers cited above, the sum of the ratios of thickness to dielectric constant for the three capacitive elements of the circuit, air gap, substrate, and insulation, is 0.151. Thus, the voltage required to start a corona is $8{,}750 \times 0.151/0.125 = 10{,}570$.

Once the corona begins and the air becomes conducting, the voltage in the air gap drops to about 500 volts, leaving 10,070 volts (10,570 – 500) to be safely applied across substrate and insulation.

At this point, there are only two capacitive elements of the circuit: substrate being treated and insulation. The sum of the ratios of thickness to dielectric constant is now 0.026. The voltage drop across the substrate becomes 387.3, or 193.7 volts per mil. For the layer of insulation, it is 9,682.7, or 78.9 volts per mil. Both values are well below the breakdown voltage of each material, assuming the substrate to be PE and the insulation to be Hypalon. In Table 2.1, breakdown voltages (dielectric strength) and dielectric constants for some film and insulation materials are shown.

EFFECTS OF TREATMENT ON WETTABILITY AND ADHESION

The effects of corona treatment may be measured in two ways. One is by measuring the effects on wetting; another is by running actual laboratory adhesion tests on the treated material. The wetting test is very rapid. Solutions of various concentrations of ethyl alcohol and water, each with a different surface tension, may be wiped on the treated surface. If the converter has an analytical balance and a platinum ring of known dimensions so that the surface tension of a liquid may be measured, it is possible to make one's own solutions of whatever combinations of liquids one desires.

A kit is available commercially of mixtures of ethyl cellosolve and formamide ranging in surface tension from 25 dynes/cm up to 50 or 60. It

TABLE 2.1. Dielectric Properties of Some Plastic Films and Insulating Materials.

Substance	Dielectric Strength (volts/mil)	Dielectric Constant
Low-density PE	450–1000	2.25–2.35
Medium-density PE	450–510	2.25–2.35
High-density PE	450–500	2.39–2.35
Ethylene-vinyl acetate (EVA)	600	2.6–3.2
Polypropylene	500–600	2.2–2.6
Polyethylene terephthalate (PET)	560	3.2–3.8
Nylon 6	440–510	3.5–4.7
Nylon 66	470	3.4–3.6
Hypalon	400	5.0–6.0
Silicone	450	4.0–5.0
Ceramic	500+	8.0–10.0

Sources: Harper, C. A., ed., *Handbook of Plastics and Elastomers.* McGraw-Hill, New York; and Simonds, H. et al. *Concise Guide to Plastics.* Reinhold Pub., New York, and Pillar Technologies, Inc., Commercial Literature.

should be remembered that wetting is a necessary, but not a sufficient, condition signifying that the level of treatment is sufficient to give adequate adhesion. Nevertheless, it is often the only test used.

Today, the market supplies treating stations with multiple electrodes, cooling apparatus which blows air through the air gap and over the assembly of electrodes and substrate carrier, and methods for exhausting the ozone formed in the air gap. Some are even made with ozone decomposition tanks. It is possible to obtain treating stations ranging from very simple to complex models satisfying a multitude of needs, but all of them do essentially the same thing – treat a substrate with a CD. Some even have methods of adjusting the width of the air gap to optimize treater conditions.

What Treatment Does

Precisely what treatment does to a surface is still the subject of much discussion and controversy. There are some things it does on which most people agree:

- It does clean off the surface. It removes adsorbed layers of water and perhaps other things with which a substrate surface may be contaminated and on this planet, almost all surfaces have a layer, even if it is only one molecule thick, of adsorbed contamination.
- It does oxidize the surface. Chemical analyses of surfaces of substrates before and after treatment indicate an increase in oxygen content through the formation of hydroxyl groups and, to a lesser extent, carbonyl groups.
- There is a strong possibility of chain scission and the formation of free radicals with potentially wide variations of molecular weight.
- It has been demonstrated repeatedly that level of treatment depends on the power applied in the air gap and the machine speed, i.e., watts/unit area/unit time.
- Different films need different levels of treatments. Film containing slip or anti-block agents will need more power applied to it to reach the same level of treatment as a film containing less or none of these agents. Treatment of polypropylene or a high- or medium-density PE may be somewhat different from treatment of a low-density PE depending on how the films are formulated.

But there is much on which there is little agreement.

There are some who contend that the effect of CD treatment is primarily caused by oxidation of the surface and the formation of polar groups, increasing wettability and presenting sites on which adhesion may take place. But it has been shown that the effect occurs in an atmosphere of helium or nitrogen. The answer provided by believers in the oxidation theory is that since helium and nitrogen are both made by liquefying air, enough oxygen is present in both (as an impurity) and enough oxygen is present in the material being treated (as an impurity) to produce oxidation on treatment.

Others hold to a ballistic theory, i.e., that the force with which the ions in the air are driven against the web is strong enough to produce microscopic dents, causing an increase in surface area and allowing an ink or adhesive to hook itself into the surface. However, this does not really serve to explain the increase in wettability. While many people agree that the bombardment may produce some declivities in the surface, they also say it is neither a major, nor the only, effect.

Next, there are the free radical theorists, who argue that ion bombardment produces chain scission and that the scission may result in varying molecular weights of the free radicals so formed.

Low molecular weight free radicals tend to interact to form larger molecular groups, and if oxidation does occur (it has been shown repeatedly that so long as treatment is performed open to the atmosphere that oxidation must occur), many of these molecules or groups of molecules will contain hydroxyl, carbonyl, perhaps ether linkages, and possibly carboxylic acid groups.

Nothing has been said about the possible formation of nitrogen groups such as imino and amino groups but among the bombarding particles there are, undoubtedly, some nitrogen ions that may also interact with the surface.

High molecular weight free radicals will be relatively immobile and tend to persist for some time. This is, perhaps, one explanation of why some treated films remain treated after months or even years in storage.

In addition to the above, there are those who hold to the electret theory. An electret is a permanently electrified body that has opposite charges at its extremities. The electrets, of course, are microscopic in size and the film surface is electrically neutral. The rationale behind this thought is that because rapid cooling of certain waxes are in an electric field results in the formation of electrets, and since waxes are hydrocarbons of high molecular weight similar to polyolefins, that a similar phenomenon occurs in polyolefin films. This would account for the relative permanence of treatment some of these films display and if one is a believer in the electrostatic theory of bonding, this becomes a logical notion.

It is quite likely that CD treatment causes all of these effects: micropitting, oxidation, free-radical formation, and the making of electrets. The situation is reminiscent of the Omar Khayyám quatrain (Fitzgerald translation):

> Myself when young, did eagerly frequent
> Doctor and Saint and heard great argument
> About it and about, but evermore
> Came out the same door where in I went.

CD as a process is deserving of much serious study. Perhaps, someone should take a long look at some of the work on plasmas that has been done by fusion researchers. Also, it would be useful to examine what has been done in the electronic chip industry, and what can be done using atmospheres of gases other than air and other temperatures. Corona discharge treatment is a startling process and accomplishes a great deal. A better understanding of what it does to a surface would probably lead to even greater accomplishments.

CHAPTER 3

Unit Operation 2 – Printing

A large part of the converter's activity is printing a legend on what will become a customer's package that tells what the product is, its ingredients, who distributes or manufactures it, the bar code identifying the product and whatever other information may be required by various government agencies. To do this economically and with minimal effort, certain equipment has been devised and used in conjunction with well-defined processes for printing.

PRINTING METHODS

There are three major methods of printing in use: relief, intaglio, and planographic. Relief printing, the oldest, is used in modern letterpress and flexographic methods. In this process, the printing areas are raised above the plane of the supporting printing plate material. Only this raised portion is inked, and transfers a part of its ink to the substrate when the inked image-bearing areas are pressed against the substrate.

Next, there is intaglio printing, in which the printing areas are below the surface and the non-printing areas are at the surface of the object doing the printing. For roll-to-roll printing, where the process is known as rotogravure, this is an engraved cylinder. This process is generally known as gravure printing, both roll and sheet-fed.

Finally, there is planographic printing, in which the printing and non-printing areas lie in the same plane. This method is used in lithography, both offset and direct. Because the printing plates used today are thin slabs of aluminum or zinc, offset lithography is used, in which the image is transferred by contact pressure to a rubber "blanket" which does the actual printing on the stock. With each cycle of the machine, the non-printing area of the plate is wetted with an aqueous solution. This permits the ink to wet only the printing area.

Offsetting is used in modifications of both letterpress and gravure printing. When it is done with letterpress, it is called *letterset,* in the case of gravure, it is simply *offset gravure.* Using these modifications of standard processes is not the result of whimsicality. There are good reasons for them, one of the major ones being that they extend the life of the printing plates, forms, cylinders, wood cuts, or whatever image-bearing object is in use.

There are other processes, such as silk screening and collotype, but these are not used extensively, particularly by converters.

FLEXOGRAPHY AND ROTOGRAVURE—THE PROCESSES OF CONVERTING

The processes used by the converter are primarily relief printing in the form of flexography and intaglio printing in the form of rotogravure. Lithography, widely favored in carton manufacture, is used for flexible packaging only rarely; an example being the process-printed envelopes for spice mixes, such as those for General Foods and R. T. French.

Rotogravure and flexography are used by the converter because both are well-suited for the printing of roll-to-roll products that are now basic to flexible packaging, the one for historic reasons, and the other because of its low cost and availability.

In the late 1920s and early 1930s, when flexible packaging was quite young, it became necessary to print on paper-backed aluminum foil and the newly developed cellophane wrappers. Some products were produced using letterpress printing, mostly sheet-fed; a slow, cumbersome method. With the repeal of the Volstead Act, paper-backed aluminum foil beer bottle labels became quite popular and a high-speed printing method became a requirement.

Reynolds Metals, the developer of the product, purchased four-color Chambon rotogravure presses in France and brought them into America. Other converters followed suit, buying presses from English, German, French, Italian, and American manufacturers.

Flexography had been introduced in the United States by the C. A. Holweg Company of France. It was used to print paper stocks in-line with a bag machine, producing a printed paper bag in a single operation. The printing plates, being made of rubber, swelled in some organic solvents, and the inks used were either water or alcohol solutions. Colors were made from various organic dyes, many of which were derived from aniline. Hence the name, *aniline printing.*

Flexography, at the time, was hardly considered suitable for the printing of either aluminum foil or cellophane. The inks bled, so it was difficult to print a sharp, clean image, and the colors had poor light-fastness. Also, no film-forming resin was used in the inks. Since most of the dyes used were basic, they were fixed, or made somewhat insoluble in water, by addition of a weak acid (usually acetic or tannic).

The process selected for handling roll-to-roll printing was rotogravure. This method continued to be used through the 1930s and into the 1940s.

With the burgeoning of converting in the post-World War II era and the wide- spread use of PE, flexography came of age. Presses were built with dryers, some with between-color dryers. The central impression press fit into the printing of PE well because it was easier to hold registration of color-on-color printing.

What made all this practical was the remarkable development in inks done by such companies as Intag, Delval, Converters Ink and others using ketone resins, SS nitrocellulose, the then novel polyamides (originally developed by the Northern Regional Research Laboratory of the Department of Agriculture, later taken over by the General Mills Chemical Co., then sold to Henkel, a large German chemical company) and other film-forming materials. The inks adhered well to PE, when properly surface treated, ran well, and, if adequately dried, did not bleed, smell, or run when wetted. The notion of drying by absorption, as in the early days of printing on paper, was abandoned since aluminum foil and cellophane were non-absorbent. And if butyl rubber plates were used, resistance to methyl ethyl ketone became sufficient to permit printing with polyvinyl chloride inks, if necessary.

Even the name of the process was changed, from aniline printing to *flexographic printing*. The movement to change the name was led by Franklin Moss, the founder of Mosstype Corp. – a manufacturer of flexographic plates, plate mounting machinery for such plates, and auxiliary equipment for flexographic printing.

This process became the one the converters chose. It was quite adequate for much of the market the converter served. A good flexographic press required a considerably lower capital investment than a rotogravure press. The cost of rubber plates was far less than the cost of engraved cylinders. Even for very long runs, where the cylinder costs were amortized over many, many impressions, flexography was less costly.

However, in the area of halftone work, flexography does not have quite the quality of rotogravure. While some advances in tone and process printing by flexography have been made, (as discussed later) both rotogravure and lithography have remained superior to it in this area. This problem is relatively minor from a marketing viewpoint since much of flexible packaging shuns tone work, relying more on bold, solid colors, which the flexographic printing process produces quite well.

Yet there is a small, but persistent interest on the part of both the converter and the packager in tone and process printing by flexography to do pictorial work less expensively than can be done with either rotogravure or lithography.

Many converters have both rotogravure and flexographic presses; some few have only rotogravure presses. The largest number, by far, of converters are flexographic specialists. To the original few manufacturers of such presses, the Potdevins, Klingroses, and Kidders, there have been added legions of press manufacturers such as FMC, Paper Converting Machinery, Dritec, Hudson-Sharp, and others, and there are many rotogravure press makers. The converter has a wide choice for purchase of new machinery, both in the U.S. and abroad.

THE BEGINNING—ORIGINAL ARTWORK

The process of taking a customer's idea of what the package is to look like and making the package look like it, is a non-simple procedure. Its inception is the idea.

First, the idea becomes a drawing of a finished package in full color done by an artist. An effort is made, by the design, to convey to the consumer a sense of what it is the package contains and to pique at least some curiosity on the part of the potential buyer. If the package design can stimulate the buyer to pick it up and turn it over, stare at it, read the ingredient copy or any other part of what the package says, it is a successful design. In the welter of packaged products displayed in any single category on supermarket shelves, there is a kind of sameness among the various brands.

Hence the use of metallized films; in addition to the added barrier properties, the gloss and brightness tend to be eye-catching. But what can be done when all the packages of a particular product are made with metallized films?

The search for a unique, easily identified package is constant. Establishing a product's brand name and fixing it in the consumer's psyche is too much to be carried by the package alone, but the package must carry part of the burden, and must fit neatly into the total marketing effort made to promote the product. This is not the business of the converter but of the customer. Some converters try to make a contribution to the package design. Others comment only when asked.

Generally, the converter accepts the design of the package as presented, and whatever changes in it the converter might suggest are usually of a technical nature, such as a change in some dimensions, to enable the artwork to fit on the printing cylinder so that when the plates are mounted on the plate cylinder, the cylinder can run at the gear pitch line.

PRODUCTION ARTWORK

The artwork, as received, cannot be translated into printing plates, without further changes being made. It is converted into what is called *production artwork.* Production artwork consists of a set of photographs of the design on cellulose acetate sheet, with each color separated out from the design, made from the original artwork.

A set of negatives of the design are made on the cellulose acetate. An artist paints out everything except one color. This is the color which is being separated from the design. Hence the term color separation. One is made for each color.

If there is no need for halftone work because the design contains no pictorial matter, and 95 percent of all flexographic package printing has no pictorial work, the color separations made by this procedure are adequate to proceed with making printing plates.

If process printing is required, the color separations are made photographically using halftone screens and color filters. A tone screen grid, a photograph of the design, and filters, of which there are three – yellow, magenta (red), and cyan (blue) – are all used in making the color separations. A screen grid, a photographic negative of the design, the proper filters to screen out the unwanted material, and a photographic plate are all inserted into a frame from which the air is exhausted. The frame has glass surfaces made of an optical glass. The air in the frame is pumped out insuring that no bubbles are present to distort the photograph. The frames are known as vacuum frames. Ordinary light is used, but it is the very intense light of a carbon arc lamp. The film emulsion is a very fine-grained silver halide type.

Up to this point, all printing processes make negatives of the production artwork this way. In turning these photographs into printing plates, the processes separate.

MAKING THE PRINTING PLATE

In flexography, the first step is making a *master mold,* as it is called. This is a form that has survival qualities so that it may be used repeatedly. From this, a mold can be made and from the mold, rubber printing plates can be made with reasonable precision.

For the master mold, a sheet of magnesium, aluminum, zinc, or copper is used. Zinc is most common, copper is used for very fine line work or halftones, and magnesium is used when there is nothing of a critical nature about the job.

The sheet of metal to be made into a master mold is coated with a photosensitive composition that does not contain silver halides and requires

UV light to harden it, such as gum arabic. Using the color separations of the production artwork, a master mold for each color is prepared, using a vacuum frame and UV light.

That part of the photosensitive coating exposed to the light hardens and becomes quite chemically resistant, the part not exposed is easily removed leaving bare metal. Etching of the metal sheet is done in a machine made expressly for this purpose. The etching solution is an acid bath, often a solution of ferric chloride containing some oil and additional agents to help control poor etching, e.g., undercutting of deep etches, and irregularities in shallow ones.

The etching solution is rinsed off and the metal sheet is dried and ready to assume its role of "master mold."

A matrix is prepared from the master mold in a compression press. Molding material consisting of a phenolic molding powder filled with cellulose or asbestos fiber is commonly formed into matrices. The identical material is also available in sheet form. The latter is ordinarily used in the preparation of matrices. The cellulose-filled material tends to shrink more than the material that is asbestos filled. Since the extent of shrinkage for both materials is known, it is easily compensated for.

The matrix, when completed, is the mold used to make the rubber plate with which the printing is to be done.

There are a variety of elastomeric materials to choose from for printing plates; natural rubber, neoprene, nitrile rubber, butyl rubber, et al. All must be compounded and filled with substances serving some purpose; some inert to lower cost, some for coloring purposes, some to add flexibility, and some to reinforce and strengthen. Some chemicals are added as antioxidants to reduce embrittlement with age, some as accelerators to speed up vulcanization to improve restitution after compression.

The rubber used most often is natural rubber with zinc oxide reinforcement and a reddish colorant, with enough sulfur added to effect a mild vulcanization during the molding operation. It is compounded to yield properties satisfactory for use as a printing plate, including good wettability by the inks to be used, Shore sclerescope hardness falling in the proper range (55–70), good resiliency, survivability through many impressions, etc. All of these properties are measurable.

Prior to molding the plate, a decision must be made as to thickness of the rubber. This will depend on the nature of the job. It is largely a judgmental decision, based on experience and an intimate knowledge of the vagaries of the press to be used. Thicknesses from 0.040 to 0.187 inches, largely in increments of 0.015 inches are available and any other thickness can be made with payment of a premium. The thickness of rubber most commonly used is 0.125 inch.

Having selected a given thickness of rubber as being suitable for the job to be printed, the thickness of the bearers is calculated. These are metal blocks placed at either end of the serving tray to prevent the upper platen of the molding press, in which the matrices are formed, from crushing the plate and the matrix. This will happen if the downward motion of the upper platen is unimpeded. The bearers provide the impediment.

The serving tray is a flat plate of steel on which the matrix, rubber molding material, and the release sheet put over the rubber to prevent its adhesion to the heated upper platen of the molding press are placed. Usually a good grade of Holland cloth is used for the release sheet. It is at each end of the serving tray that the bearers are placed. Bearer thickness is calculated by adding together desired thickness of rubber plate, thickness of bottom of matrix, thickness of cover sheet, and a compensating small amount (on the order of 0.0005 inches) for shrinkage. All of this having been done, the plates are molded.

The upper platen is held at about 300°F and the rubber flows under a pressure that may be as high as 500–600 psi, filling the matrix. When molding is complete, the temperature of the platen is raised to about 375°F and held there for around 10 minutes to vulcanize the plate mildly.

At every step of the way, from artwork to molded plate, it is necessary to check results carefully. Photographs may need retouching, engravings may need redoing or some minor correction, and the matrices of the individual colors may need to be molded more than once. Frequently, the rubber plates require some small amount of grinding after mounting to make them fit the pitch diameter of the gear that turns the cylinders on which the plates are mounted. This is done with the cylinder in the press.

Platemaking, despite the use of some rather advanced equipment, is an art, and the variation in talent from one platemaker to another may be quite startling.

In recent years, there has been commercial development of photopolymers, some of which are quite suitable as plates. Using the negative of each color separation, a positive image is made on a sheet of photopolymer with UV light. The parts exposed to the UV harden, the parts not exposed remain soft and are readily removed.

Such a device can be used as a printing plate eliminating the steps of master mold formation, matrix making, and molding the rubber plate. DuPont's Dycril is one such material that has had some success. However, many converters prefer rubber plates.

Platemaking is not confined to a separate platemaking industry. Many converters make their own rubber plates. Some, who are rotogravure specialists, even make their own rotogravure cylinders.

To summarize the process: first, several photographic negatives are made on emulsion-coated cellulose acetate. Next, an artist paints out the unwanted portion of the negative to leave only one color. This is done for each color. Then, a photograph of each negative is made on a photosensitive material-coated metal sheet, usually zinc. The zinc sheet is etched, washed, and dried. This metal sheet is the master plate and is a positive representation of the final plates. A phenolic resin matrix is made from the etched metal form, providing a negative mold. Finally, a rubber printing plate is made using the matrix as a mold. This plate will be used for printing.

This completes what is, essentially, only the first step, but it is a major step. It is all but impossible to produce a satisfactory job without reasonably well-made printing plates. Fine printing, albeit of simple design, has produced great pride among the converting fraternity. Some major awards are given by the technical and trade associations of both the printing and the flexible packaging industries for quality printing by converters.

MOUNTING THE PLATES

Having made the plates, the next step is to mount them on the plate cylinders with all colors in register. To do this, mechanical assistance is necessary. The first flexographic mounting and proofing machine was devised in the early 1940s by Franklin Moss of Mosstype Corp., who made an adaptation of the letterpress line-up table.

On the plate-mounting machine, the printing cylinder with bearings on its journals is fitted into V-blocks. A mechanism for locking it in place is provided. Above the cylinder a calibrated measuring straightedge is placed. A dividing head, such as that on a lathe, is used for measuring distance around the circumference of the roll. A scriber pen for marking is incorporated into the apparatus. The calibrated straightedge, perfectly parallel to the cylinder axis and equidistant from it along its entire length, measures horizontal spacing, and the dividing head measures the extent to which the cylinder need be rotated for exact location of a scribe line on a plate.

Earle Harley, then connected with H. H. Heinrich, a press manufacturer, altered the equipment by replacing mechanical measuring devices with optical ones. Over the years, since Moss originated the machine, it has gone through many modifications and evolved into today's precision apparatus without any radical change in the principles on which the original was based. Today, both the Moss and Harley mounter types are employed.

In mounting, the first step is to wind a sheet of proofing paper around the impression cylinder, mounted in front of the printing roll in Moss's equipment and below it in the Harley modification, carefully lining it up with the scribe line close to the edge of the roll. The paper is made to butt join at its ends and is held to the roll with pressure-sensitive tape (most likely, masking tape). On this sheet of proofing paper, using the tools provided on the machine, a sketch is made of the exact layout of the plates on the cylinder, staggering them to distribute pressure uniformly, and providing all the needed information as to leading and trailing edges, eye-spot location, and any other information regarding the work.

A sheet of *stickyback* is put around the plate cylinder carefully pressing out any air bubbles and without unduly stretching it. The protective covering is carefully removed, strip by strip and with a grease pencil the layout sketch is drawn on the stickyback to whatever extent is needed to mount the plates with precision.

The plates are then pressed against the stickyback carefully, without stretching them, so that the scribe lines on each plate correspond to scribe lines on the stickyback, which, in turn, correspond to scribe lines on the layout drawing. The entire procedure should be done unhurriedly and with care. Cleanliness is, obviously, essential. Should the stickyback not be very sticky, it can be activated with any hydrocarbon solvent (toluene is a good one) by gently wiping solvent on the surface, then permitting it to air dry.

After the plates are mounted on each plate cylinder, and the cylinders placed in the press, they may be very lightly ground in place if there are any small hills or valleys.

AND NOW—TO PRINT

Once the plate cylinders are mounted in the press with the proper gears, the substrate to be printed is threaded through the machine, under the between-color driers, between the printing plates and the impression roll, into the oven where the inks are dried, and then to the rewind where it is wound up as a finished, dry, clean, printed roll.

There are two major designs of machine for accomplishing this result: the stack press, shown schematically in Figure 3.1, and the central (or single) impression machine shown in Figure 3.2. The stack press has the color stations mounted, one above the other up one side, and one below the other down the other side of the frame of the press. The central impression press has a large central cylinder around which the color stations are mounted. The central cylinder acts as the impression roll for all the colors, hence its name.

A single color station illustrating the typical two-roll ink distribution system of flexography is shown in Figure 3.3. A roll turning in the ink pan (rubber covered) picks up the ink and transfers it by pressure to a steel roll engraved with an overall pattern of cells which hold the ink. This is the

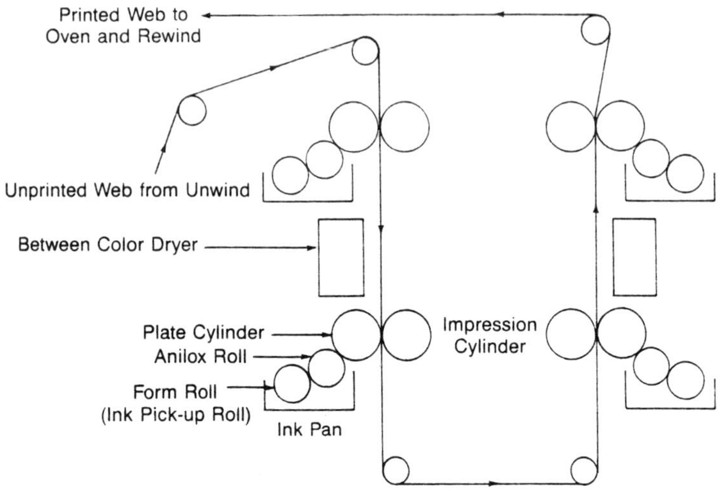

FIGURE 3.1. Stack flexographic press (schematic).

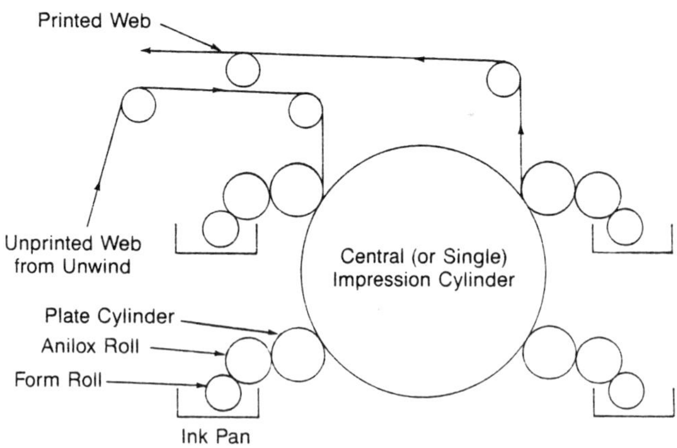

Between color dryers are not shown on central impression
press because the drawing is quite crowded.

FIGURE 3.2. Central impression press (schematic). Between-color dryers are not shown on central impression press because the drawing is quite crowded.

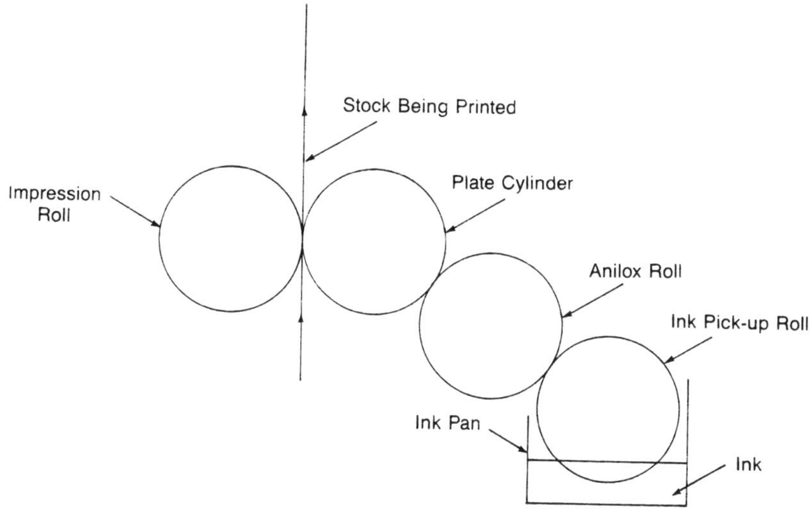

FIGURE 3.3. Standard two-roll flexographic ink distribution system.

anilox roll and the cell pattern may vary. The anilox roll transfers ink by contact and pressure to the plate cylinder on which the rubber printing plates are mounted. The stock to be printed runs between this roll and another called the impression roll, which squeezes the stock against the plate, causing transfer of ink.

On a stack press, each color station has its own impression roll. On a central impression press, the stock runs between each plate cylinder and the single impression roll. The plates of each color are pressed against the stock, which, in turn, is pressed against the impression cylinder. Both methods are readily understood systems of image transfer using fluid inks.

Currently, there are many well-designed machines available for flexographic printing, ranging from a stack press with four-color stations (Wolverine) to central impression eight-color equipment specifically designed for the printing of bread bags (Hudson-Sharp, FMC). Among converters, the equipment and procedures for high-quality printing are truly at hand.

THE TWO-ROLL INK DISTRIBUTION SYSTEM

In flexographic printing, ink distribution is essentially a two-roll system, shown schematically in Figure 3.3. The fountain roll picks up the ink and transfers it by contact pressure to the anilox roll. The anilox roll transfers ink by contact pressure to the plates mounted on the plate cylinder. Whatever ink is on the plates is the maximum amount of ink that may be transferred to the web.

How much is transferred is dependent on the interfacial tensions between the ink and the various solid objects wetted by it. At each stage, the ink film divides between the two objects being pressed together: between fountain roll and anilox roll, between anilox roll and plate cylinder, and between plates and stock. During any single run, since interfacial tensions do not vary provided solvent content of ink is held reasonably constant, they do not affect consistency of quantity of ink deposited.

It has been found experimentally that ink deposited varies with machine speed. But this is only one factor affecting ink laydown. There are others:

- nip pressure
  - fountain roll/anilox roll
  - anilox roll/plate cylinder
  - plate cylinder/stock
- ink viscosity
- roll diameter (or radius)
- hardness of rubber
- cell volume of anilox roll
- machine (or roll) speed

Not all of these affect the thickness of the ink layer equally. The amount of ink deposited does vary strongly with the speed of the machine. To lay down a constant thickness of ink, it is important to hold all these variables constant during a run.

A partially theoretical, partially empirical relation among all these variables may be expressed as:

$$h^2 + Prh/K - PrnV/4K^{2.68} = 0 \quad (3.1)$$

where

h = thickness of ink film (inches)
P = nip pressure (psi)
K = bulk modulus of rubber at given hardness (psi)
r = mean roll radius (inches)
μ = ink viscosity (lbs. sec./square inch)
V = machine speed (inches/sec.)

For illustrative purposes only, the conditions taken, on which the curves of Figures 3.4 to 3.7 are based are:

P = 120 psi
K = 284,000 psi, a reasonable value for a medium hard rubber
r = 3.5 inches
μ = 100 cps = 1 poise, relatively high for an ink, but easy to calculate with

In using Equation (3.1) to calculate values for the plotted curves, the anilox roll was treated as if it were a smooth roll. This is only partially true. While it is run with a bead along its length, it is engraved. Thus it acts both as a smooth roll and an engraved one. This may account for the rather strange exponent of 2.68 in the last term. No assumption was made as to how the ink divides between the sets of contacting surfaces, fountain roll/anilox, anilox/plate cylinder, plate cylinder/stock. In any case, with all other conditions fixed, this will not vary during a run.

The curves in Figures 3.4 to 3.7 show wet film thickness, a measure of ink laydown, plotted against machine speed for two different sets of values for P, K, r, and n. The curves show how each of these variables affects quantity of ink deposited. A change in roll radius or a change in nip pressure affects ink laydown less than a change in rubber hardness or ink viscosity.

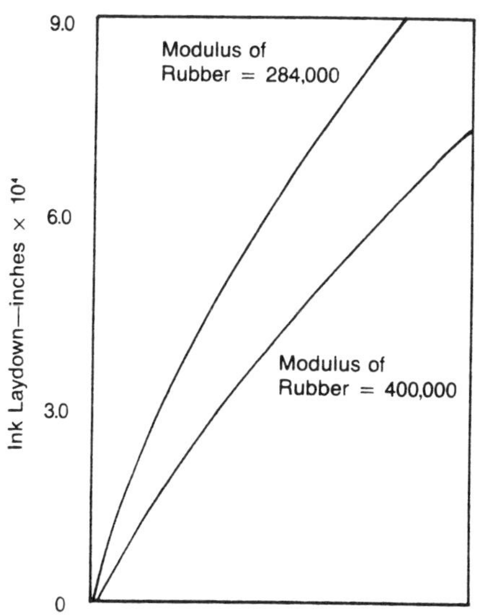

FIGURE 3.4. Effect of rubber hardness on ink laydown.

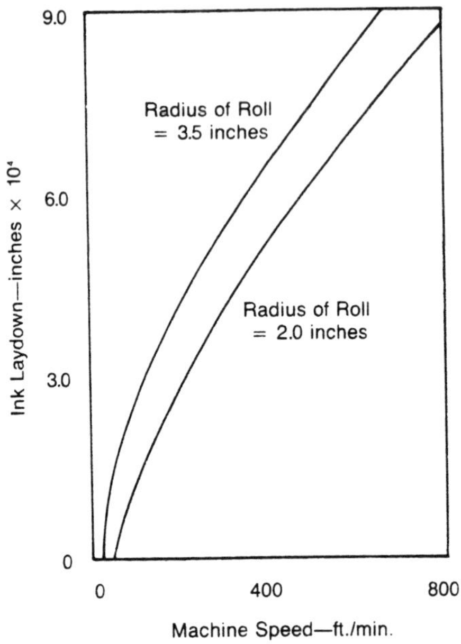

FIGURE 3.5. Effect of roll radius on ink laydown.

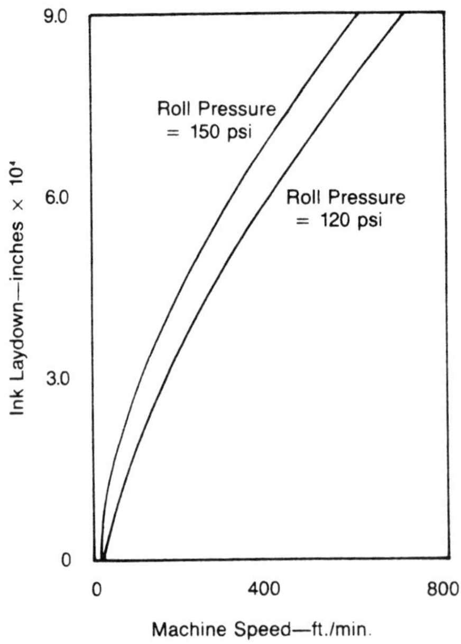

FIGURE 3.6. Effect of roll pressure on ink laydown.

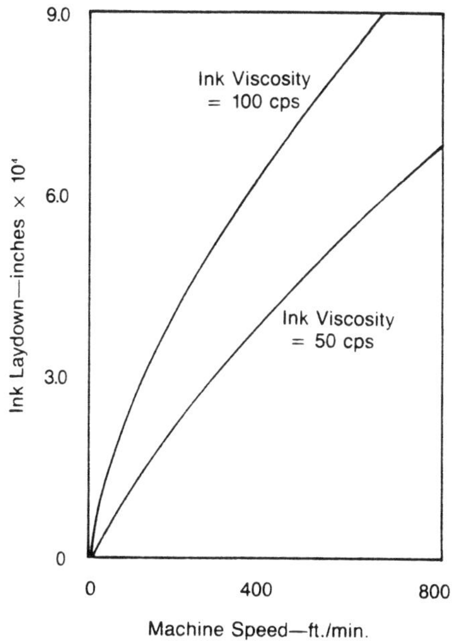

FIGURE 3.7. Effect of ink viscosity on ink laydown.

But all these changes in ink quantity deposited are minor compared with the extent laydown is altered with variations in machine speed.

While Equation (3.1) is not exact, it does show, in semi-quantitative terms how some factors affect ink laydown in the conventional flexographic ink-metering system. Fortunately, for solid colors, this does not seriously influence job quality.

PICTORIAL PRINTING WITH RUBBER PLATES

In the last two decades, serious efforts have been put into doing tone and process printing with rubber plates and some rather creditable pictorial jobs have been produced. When doing this, there are some caveats and special procedures to be followed.

It has been found that copper is the best material from which to make master molds for halftones. Copper sheet can be consistently rolled with a finer grain size than zinc, aluminum, or magnesium, making quite precise engravings possible.

The tone screens used are in the range of 90 to 120 lines per inch although there has been some use of screens as fine as 150 lines per inch and as coarse as 65. The rubber dots are slightly smaller than those normally etched in metal because of the tendency of rubber to spread even under very light pressure.

If care is not exercised in platemaking for tone work, two unwanted effects may result: first, the line of dots may be obvious to the naked eye, and, second, a moire pattern may develop.

These two tendencies may be overcome if the grid of the halftone screen for each color makes a different angle with the horizontal. From an analysis according to the physical laws of optics, the least visible line of dots is one at a 45° angle. If a single-tone plate were needed, the angle selected for orientation of the grid pattern of the halftone screen would be 45°.

When four-color process work is done (yellow, magenta, cyan, and black), the dominant color, often the cyan (blue), is set with its halftone grid angled 45° to the horizontal plane. Then the orientation of the yellow is set at 75°, that of the magenta, 105°, and that of the black, 135°. Thus, no one plate has its dots following a line parallel to any other plate.

While there is agreement among those skilled in the art on the need for the dominant color to have its halftone grid oriented at 45° to the horizontal, there is little agreement as to how to orient the other colors, except that their orientations should all be different. The grid itself consists of two sets of straight lines perpendicular to each other. What is referred to here is the orientation of the entire grid pattern.

Tone and process plates need to be molded to a finer tolerance than those for solid and line work. Usually, they are molded to within 0.001″ of desired thickness and ground to 0.0005″ prior to mounting using the identical rubber thickness to start of 0.125″. For tone printing on film (polyethylene, polypropylene, polyester), a plate hardness of 60–70 Shore Durometer A is generally suggested, a somewhat higher hardness than the 50–55 in use for solids and line work. All of this is done to offset the tendency to spread of the rubber dots.

Concentricity, rigidity and general mechanical properties of the plate cylinders are more important in tone and process work than in solids and line work. Total indicated runout (T.I.R.) of the plate cylinder should be held to 0.001″ and certainly not exceed 0.002″ if reasonable printing quality with tone and process plates is to be maintained.

Plates should be mounted to print at the gear pitch line when possible. Most flexographic plate mounters tend to mount plates above the gear pitch line, causing the rubber to spread when contact is made with the stock being printed. This is not a serious factor in solid and line work, but in tone and process work it leads to spreading of the halftone dots, too large a dot impression, and a loss of fidelity of reproduction. Printing of halftone plates with process colors should be done with minimal pressure between fountain roll and anilox roll and between anilox roll and plate cylinder to control quantity of ink transferred as closely as possible in moving the ink from the pictorial image to the substrate being printed.

For faithful reproduction of pictorial matter by halftone plates and process colors, the amount of ink deposited must provide color matching closely the amount of color displayed in the artwork. The standard 165 or 180 pyramid cell anilox roll applies too much ink, which increases with machine speed as noted above. Even with 200- or 220-line anilox rolls, ink laydown is excessive.

Techniques have been developed for making finely engraved rolls with as many as 400 – 500 lines per inch with as carefully controlled cell depth as control of the number of lines. A 350-line cylinder is commonplace. These finely engraved rolls pick up and deposit fine layers of ink, enabling the press crew to match the ink laydown to the color in the artwork, within reasonable limits. Additionally, although ink deposited increases with press speed, the increase is not so high that economical press speeds cannot be reached without losing fidelity of reproduction of the original artwork.

Faithful reproduction of pictorial matter using the standard two-roll ink distribution system can be done at moderate press speeds if ink viscosity can be held very closely, machine speeds are kept constant, and if roll pressure is kept fixed. A skillful press crew can probably do a creditable job at press speeds of 300 – 400 fpm by exercising care.

THE REVERSE-ANGLE DOCTOR BLADE

The final touch, making good tone and process work feasible by flexography, has been the use of the reverse-angle doctor blade to remove excess ink from the anilox roll and thus make ink quantity printed independent of press speed. With this device, the maximum amount of ink that can be laid down on the substrate being printed cannot exceed the volume held in the cells of the anilox cylinder. With a fixed interfacial tension of ink and fountain roll, ink and anilox roll, ink and plates, and ink and stock being printed, the ink film's division between them will be fixed and will not vary with speed of the press.

Thus, with care exercised in platemaking, using the procedures listed, the finely engraved anilox roll, and the reverse-angle doctor blade, a quite respectable halftone print with process colors becomes a reality. The reverse-angle doctor blade is shown, schematically, in Figure 3.8.

But rubber dots do spread even with all these precautions. A composite plate material has been made consisting of about 0.100″ of rubber combined

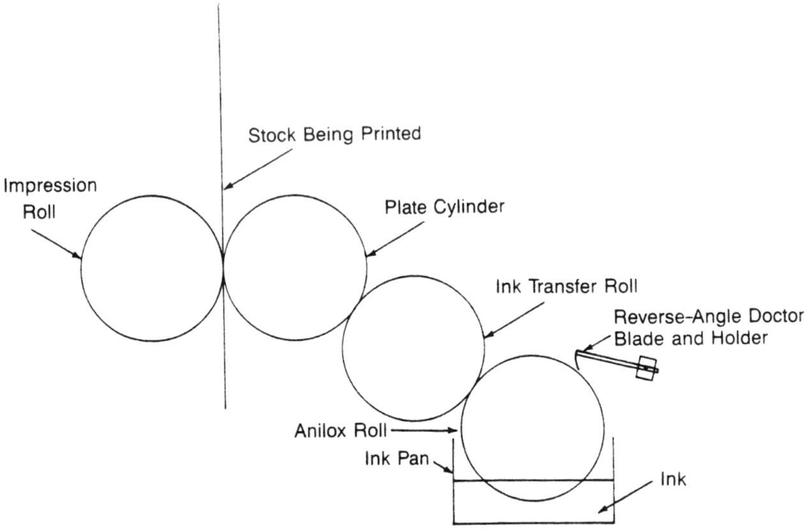

FIGURE 3.8. Reverse-angle doctor blade adaptation to flexographic ink distribution system.

with a backing of very dense foam that permits some slightly excessive pressure (as when plates are mounted above the gear pitch line) to be absorbed by the foam. Whether these plates can retain their coefficient of restitution through the compression/relaxation cycle on the press for lengthy runs is a question yet to be answered.

The reverse-angle doctor blade is positioned in a way that enables it to scrape the excess ink off the engraved cylinder, leaving only that amount of ink that can be held by the cells, analogous to a rotogravure coating situation. The force exerted by the blade on the anilox roll is of the order of oz/in^2 in contrast to the lbs/in^2 required in the standard two-roll system. Such an ink metering system makes possible the use of halftone screens with a grid of 150 lines/inch.

Utilizing all the precautions mentioned, and using the reverse-angle doctor blade, tone and process work with rubber plates becomes possible. Whether it will become a serious competitor to rotogravure or lithography is questionable. But if flexography can maintain its economic advantages over rotogravure and lithography, in some cases it may succeed in taking jobs away from the other processes.

Nonetheless, there remains the fundamental disadvantage. Rubber dots spread, metal dots do not.

CHAPTER 4

Unit Operation 3 – Coating

THERE are many ways to apply a coating to a substrate.

There are many ways to coat a wet coating or apply a wet adhesive to any moving continuous web of any material capable of forming a web. And since, before lamination, a coating of adhesive must be applied, the coating operation assumes additional importance because another operation depends on it.

COATING WITH A BLADE

One of the oldest coating methods is knife coating. Figure 4.1 shows what has been called knife over roll coating. A puddle of coating is placed on the substrate behind the knife. At each edge of the substrate a deckle or dam is attached to the frame of the machine to prevent the coating from spilling over the edge of the stock. Substrates to be coated by this process must be reasonably heavy, perhaps even thin sheets of metal. The process is used to apply a relatively thick coating (0.001″ thick or greater, dry). Many such machines were used to spread very viscous, high solids rubber coatings on fabric for tents, tarpaulins, and the like. Hence the name, *spreaders*.

The term rubber includes any elastomer: natural rubber, GR-S, nitrile rubber, neoprene, or others. Today, some of these have been replaced by isocyanate/polyol combinations or isocyanate/polyamine combinations.

While knife coating is a process rarely, if ever, used by flexible packaging converters, it has characteristics which can be observed in all coating processes.

Referring to Figure 4.1, there is no force pushing the coating through the gap between the knife edge (a rounded one having a measured radius) and the stock being coated. The gap is carefully measured and the knife tightened in position before coating is begun. The size of this gap is the primary factor affecting the coating thickness.

When the machine starts and the web begins to move, a boundary layer of coating material moves with it. The boundary layer, therefore, moves with the velocity of the web, or, at machine speeds (Prandtl's *no slip* condition). The layer of fluid above the boundary layer is dragged along with the boundary layer because of viscous forces exerted on it by the boundary layer. But there is a slight lowering of velocity because, in the transfer of kinetic energy from one layer to another, some energy is lost to friction, as one layer of fluid is pulled by the adjacent layer below it.

This loss of velocity (because of loss of energy) occurs throughout the entire coating right up to the knife edge, where the velocity of the coating is zero. Assuming a Newtonian fluid (i.e., one that obeys Newton's definition of viscosity), the equation of motion of the fluid becomes, simply:

$$v = V(1 - y/h) \tag{4.1}$$

where:

V = machine speed
v = coating velocity at y
h = distance from knife edge to top surface of web being coated
y = vertical axis

The origin is taken as being the intersection between an exact vertical extension of the knife and the horizontal line representing the direction of motion of the coated web. The x-axis is this horizontal line.

The average velocity of the coating is found to be $v_{ave} = V/2$. If Q were to denote volume of coating, then $Q = V/2 \cdot Wh = VWh/2$, where V is

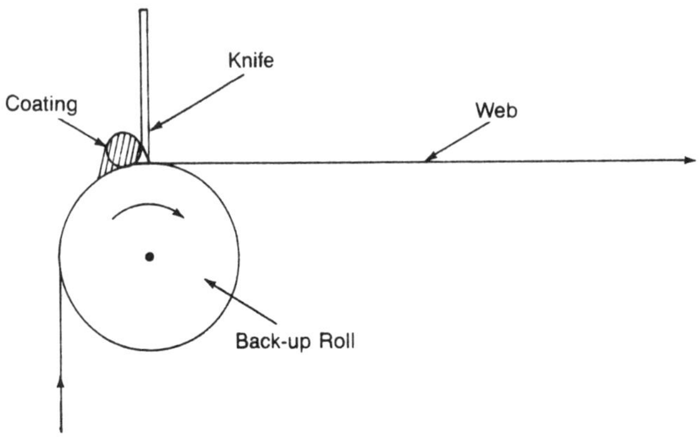

FIGURE 4.1. Knife over roll.

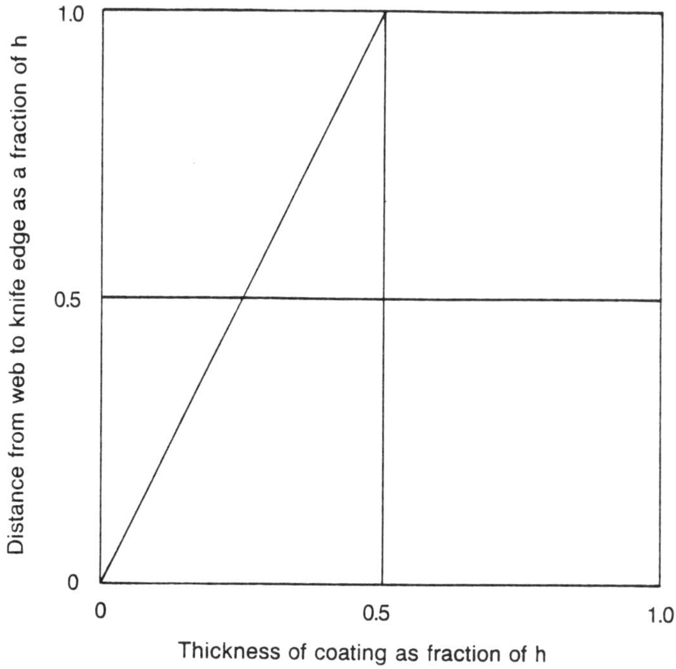

FIGURE 4.2. Thickness of coating vs. distance from web surface to knife blade.

machine speed. As indicated above, W is width of coating and thickness of coating is $h/2$.

Figure 4.2 shows the variation of coating velocity with position from the knife edge to the top surface of the coated stock. It is a straight line, which is a reasonable expectation when the physics of the situation are considered. There are no forces other than viscous forces acting on the fluid. The force retarding flow at the knife edge equals the force advancing flow at the top surface of the web being coated, both being subjected to Prandtl's no-slip condition.

DRAWDOWN

The coating thickness is less than the knife setting. It is fixed at $0.5h$ regardless of speed of machine or coating viscosity.

What is happening is that the coating is being drawn down, or stretched in the direction of motion of the web at the instant of application, because of the retarding force exerted by the knife edge.

This is true of all methods of coating, even rotogravure coating, where

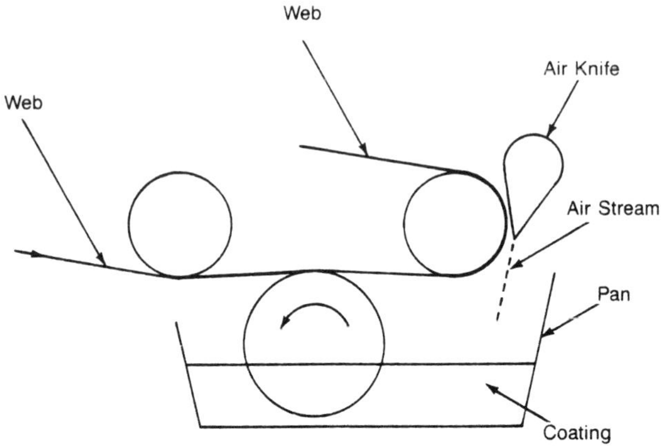

FIGURE 4.3. Air knife.

the effect of drawdown is the formation of longitudinal streaks which are sometimes extremely difficult to see. In many cases, the effect of drawdown is all but immeasurable. In the case of the example selected, it is quite significant.

The phenomenon of drawing down a coating is independent of method of application, independent of coating viscosity, and independent of nature or thickness of web. It is built into the act of transferring the coating to the substrate from the coating mechanism. It can be compensated for by applying an excess and metering the coating soon after application by some metering device, such as a wire-wound rod.

OTHER BLADE-METERING DEVICES

There is a type of knife other than the mechanical one, called an *air knife*. It is not a coating method but a metering device. The coating is applied by running the web over a roll (reverse or direct) and the air knife blows off the excess. It is useful for applying coatings of low mechanical shear stability, because in both the application and the metering, only a small amount of mechanical shear stress is encountered, unless the coating material is subjected to constant circulation. A particular use of this method is coating polyvinylidene chloride (PVdC) water dispersions on glassine or other highly calendered paper stocks. It is noisy, difficult to adjust, is really suited only to quite long runs and is not held in high regard by the converting industry. It is illustrated in Figure 4.3.

Another kind of metering device is the wire-wound rod, available in many different diameters of wound wire. When pressed against a wet, newly applied coating, it permits only a limited amount of fluid to pass through the open space between successive loops of wire. Variations in the diameters of wire result in variations in the quantity of fluid passing through; thus, different thicknesses of coating will be applied. The range of rods available runs from those with very fine wire to ones with very coarse wire. The rod is mounted on the machine and rotated either reverse to the web direction or in the same direction depending on the nature, viscosity, and quantity of coating to be metered.

They are not popular as a production mechanism among converters, but 12 – 15″ long wire wound rods are very useful in the laboratory evaluation of adhesives and coatings.

ROLL UPON ROLL

A number of coating methods are based on various roll arrangements. A simple one is shown in Figure 4.4. The rolls are placed one above the other, the two shafts parallel to each other, and the journals in a vertical line. The lower roll picks up coating from the pan and transfers a part of it to the upper roll through a nip. One roll may be rubber covered, the other chromium-plated steel. Both rolls may be metal, but this would require that they be precision ground with a total indicated runout (TIR) of 0.001″. When steel and rubber-covered rolls are used, customarily it is the steel roll which runs in the pan and picks up the coating.

This puts an immediate limitation on solvents that can be used, since swelling of the rubber covering must be avoided.

When a steel and rubber roll are used, frequently neoprene or nitrile

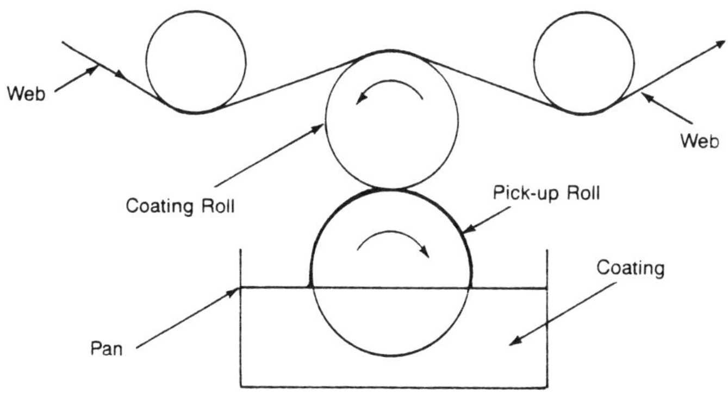

FIGURE 4.4. Roll upon roll.

rubber is the rubber chosen because of their slightly superior solvent resistance. Even when the steel roll is used as the pickup roll, the rubber roll is still thoroughly wetted by the coating, and swelling of rubber due to solvent is still possible.

How a coating divides between two objects making contact with it is important in this or any other coating mechanism. Thus far, only viscous forces play a part. As a prerequisite of being able to coat at all, all the parts of the coating mechanism and the substrate itself must be wet by the coating.

CONTACT ANGLE—A MEASURE OF WETTABILITY

It has been known for some time, that if a drop of liquid containing no impurities is placed on a solid surface, the drop often forms an angle with the surface. This is known as the *contact angle.* And for any given pair of liquid and solid, the contact angle will be a constant. Contact angles range in size from very large to zero. When the angle is zero, the liquid wets the surface and will spread over it.

W. A. Zisman and his co-workers at the U.S. Naval Research Laboratory found in the course of making studies of lubrication that if the contact angles of a series of homologous liquids, e.g., the normal alkanes, were measured and the surface tensions of the liquids were known, then a plot of cosine (contact angle) vs. liquid vapor surface tension (ST_{LV}) produced a straight line from very large angles to zero, or wetting. The surface tension at cosine (contact angle) = 1 (contact angle = 0), was called the critical surface tension.

If a liquid's surface tension exceeds this value, it does not wet the solid; if below, it does. Zisman and his co-workers also found that many non-homologous liquids also fell on the straight line or very nearly so. A typical plot of cosine (contact angle) vs. surface tension is shown in Figure 4.5. Some of the critical surface tensions determined are shown in Table 4.1.

Such data are valuable in revealing structural differences among polymers. Observe the difference in critical surface tension between polytetrafluoroethylene and polytrifluoromonochloroethylene. The difference is due to one chlorine atom in the monomers from which the polymers are made.

Using the concept of critical surface tension, it is possible to develop an idea of how a coating might divide between two surfaces where there are no forces acting on the coating other than viscous and surface ones. If it is considered that the critical surface tension of surface 1 is c_1 and that of surface 2 is c_2 then the total available active surface forces are $c_1 + c_2$. The fraction of total coating adhering to surface 1 is $c_1/(c_1 + c_2)$ and the fraction

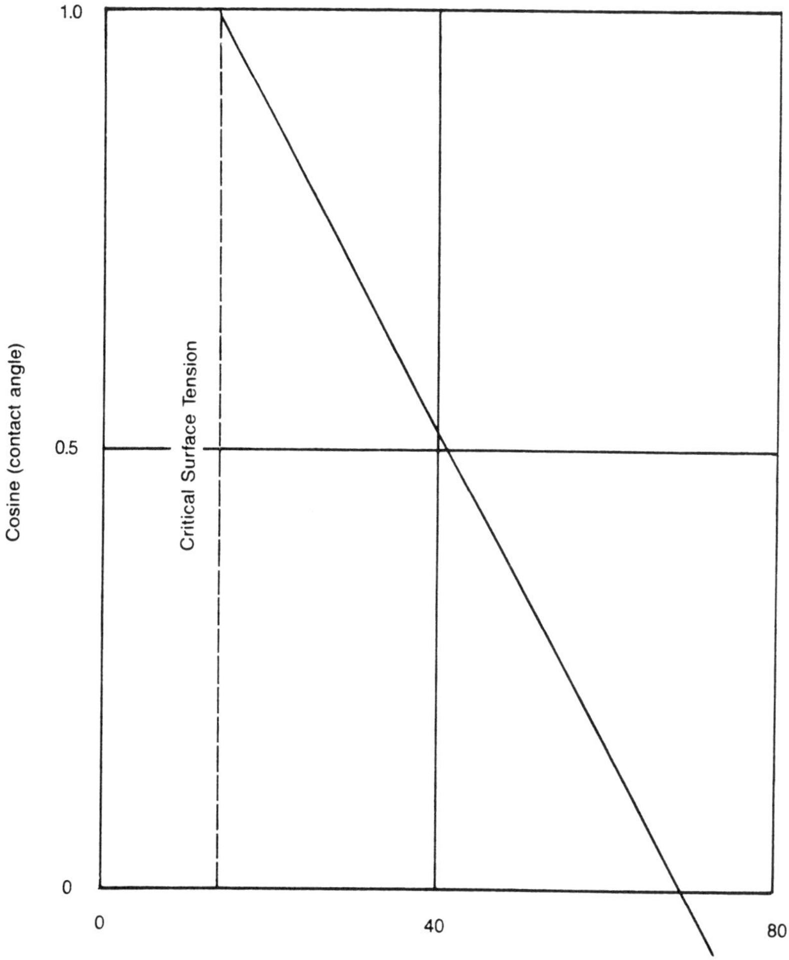

FIGURE 4.5. Cosine (contact angle) vs. surface tension.

TABLE 4.1. Critical Surface Tension of Various Polymeric Film Formers.

Polymer	Critical Surface Tension (dynes per cm)
Polyhexafluoropropylene	16.2
Polytetrafluoroethylene (Teflon)	18.5
Polyethylene (PE)	31
Polytrifluoromonochloroethylene	31
Poly(vinyl chloride) (PVC)	39
Poly(vinylidene chloride) (PVdC)	40
Poly(ethylene terephthalate) (PET)	43
Poly(hexamethylene adipamide) (Nylon 66)	46

Source: "Contact Angle, Wettability and Adhesion," *Advances in Chemistry Series*, Number 43, American Chemical Society, 1964.

of total coating adhering to surface 2 is $c_2/(c_1 + c_2)$. If these fractions are each multiplied by h (the total thickness of coating), it is possible to calculate how a coating divides between the two contacting surfaces.

The postulated calculation is based on the assumption that it is possible to measure a critical surface tension for all surfaces. However, in the determination of critical surface tensions of high-energy surfaces, difficulties have arisen because of the tendency of polar/non-polar impurities to form monolayers on high-energy surfaces. Many of these monolayers defy wetting by anything, thus causing almost all liquids to form contact angles greater than zero. One would expect a high energy, clean surface such as a metal one to be readily wet by many liquids. This is not the case, making values of critical surface tensions very difficult to obtain.

There are liquids from which a monolayer will form, where the critical surface tension of the monolayer is lower than that of the liquid itself. These are called autophobic liquids and obviously must be avoided or used sparingly in the formulation of coatings. Examples of such are 1-octanol and ethylhexanoic acid. In the formulation of water-borne materials, care should be exercised not to include as wetting agents some salts of saturated fatty acids, such as stearates, which tend to form non-wetting monolayers on high-energy surfaces.

A POSSIBLE METHOD OF CALCULATING APPLIED COATING

In order to insure that more coating will be transferred than is retained, it is necessary to apply some outside force, such as running a transfer roll faster than the web. This leads to an additional coating thickness of:

$$h_1 = [v_g - v_W]R/64\ [\pi]^2 \tag{4.2}$$

If the roll is run in reverse to the web, increasing the relative velocity between the roll and the web, then Equation (4.2) becomes

$$h_2 = [v_g - v_W]R/32\ [\pi]^2 \tag{4.3}$$

where:

h_1, h_2 = additional thickness of coating due to external force generated by difference in velocity between web and applicator roll
v_g = speed of roll
v_W = speed of web
R = radius of roll

In Equations (4.2) and (4.3), the units are in inches, seconds, and pounds, and if the acceleration due to gravity is needed, it is 386.4 inches per second^2.

The equation used for computing h, the coating thickness, is Equation (3.1) where factors such as rubber hardness, roll pressure, and coating viscosity are taken into consideration. Equations (4.2) and (4.3) are for the additional coating thickness due to different speeds of web and applicator roll.

The method shown in Figure 4.4 is also quite old, having been devised when most of the coatings in use were nitrocellulose-based and solvents that swelled rubber were used sparingly in coating formulations.

The coating roll may turn opposite to the web direction or in the same direction. If it turns in the opposite direction, it is called reverse roll coating. If in the same direction, it is called direct roll coating. When run as a reverse roll coater, in order to get complete coverage of the substrate being coated, it is necessary that the coating roll turn about 10–20% faster than the web is moving. The pressure of one roll on the other can be varied, exerting some control over the amount of coating being picked up.

Even using Equations (4.2) and (4.3) and utilizing the surface tension division between roll and substrate, predicting the coating thickness is difficult. What number does one use for the critical surface tension of the rolls? If rubber covered, what is the composition of the rubber, and is its critical surface tension known or must it be determined? If steel, is a critical surface tension measurable and if not, what value shall be used?

The amount of coating applied depends strongly on web tension, how the coating divides between pickup roll and coating roll, coating roll and substrate, coating viscosity, machine speed, and other factors. Historically, this was one of the earliest coating methods used by flexible packaging converters. There may be some machines equipped to coat this way extant today. However, because coating pickup is very dependent on web tension, which may vary during a single run, this process is not suggested as being very useful in today's markets where high performance and consistency of properties are required in most successful packaging materials. Holding the applied coating constant is difficult. This is true whether the coating roll is rotated either forward or backward, i.e., whether the machine is run as a "direct roll" or "reverse roll" coater.

A SPECIAL KIND OF REVERSE ROLL COATER

In Figure 4.6 a diagram is shown of a reverse roll coater developed many years ago by the John Waldron Co. and manufactured today by the Waldron-Hartig division of the Midland-Ross Corp. It is more complex than other reverse roll coaters, and more precise in how much coating it will lay down.

As shown in the diagram, the coating is placed in the gap between the doctor roll and the coating roll. Both of these are made of hard chromium-plated steel. The rolls are precision ground with a TIR of 0.001 inch or less. If a variable-speed device is placed between the two rolls so that the doctor roll's speed relative to the coating roll speed can be varied, the machine will be able to coat a wide range of viscosities. For low viscosities, the doctor roll will rotate very rapidly (e.g., for water dispersions). For heavy, viscous materials, the doctor roll would be rotated slowly. There have been cases

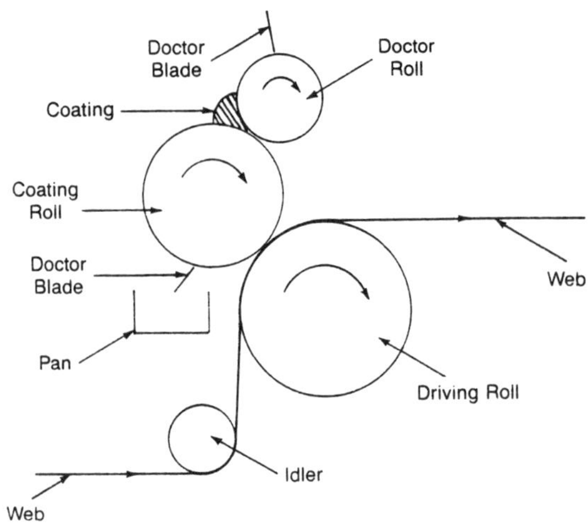

FIGURE 4.6. Waldron-Hartig reverse roll coater.

where the viscosity of the coating was so high that the driving gear was removed from the doctor roll and the machine was run as if it were a knife coater.

The gap between the doctor roll and the coating roll can be varied, thus effecting control over the coating weight. The doctor blade on the doctor roll removes whatever coating has been carried up by the roll while running. This material returns to the puddle of coating.

The gap setting is done accurately to permit only precise amounts of coating to be carried through it by the coating roll. How much coating is determined by the requirements of the job being run. At each end of the roll pair (doctor and coating) a deckle or dam is placed to prevent coating from leaking out of the puddle. The doctor blade on the coating roll removes material from the coating roll which has not been transferred to the web being coated.

In operation, the driving roll, a large-diameter steel roll covered with a layer of rubber about an inch thick is moved upward to force the web to make contact with the coating roll. As the web moves, the coating roll rotates in a direction opposite to that of the web motion. Hence the name, *reverse roll coating.*

Because of the reverse direction of the coating roll, there is a *wiping action* by the roll on the web and the tendency to draw down the coating is minimized. Customarily, the coating roll runs at a speed about 15% faster than that of the web to reduce the possibility of gaps in the coating. The quality of the coating can be seen on the exposed face of the coating roll. Whatever defect in the coating is observed will transfer to the web: holes, streaks, lumps, etc. It thus becomes possible to make some adjustments early in the run and correct the defects. In this design of reverse roll coater, the amount of coating applied is independent of web tension or variation in web thickness. Its limitations lie in the inability to apply thin coatings of less than 0.001 inch. In the range of 0.001 – 0.005 inches, this method is all but ideal. The dimensions mentioned refer to wet coating thickness.

ANOTHER TYPE OF SMOOTH ROLL COATING

One further method of using smooth rolls as coating devices is illustrated in Figure 4.7. Roll configuration is identical to that used for gravure coating, but a smooth roll without a doctor blade is used as the pickup and application roll. The impression roll is brought down and pressure applied as if coating were being done with an engraved cylinder. It is advisable to use a smoothing bar (not shown in the diagram) following the coating application to smooth out longitudinal streaks, which are common when using this

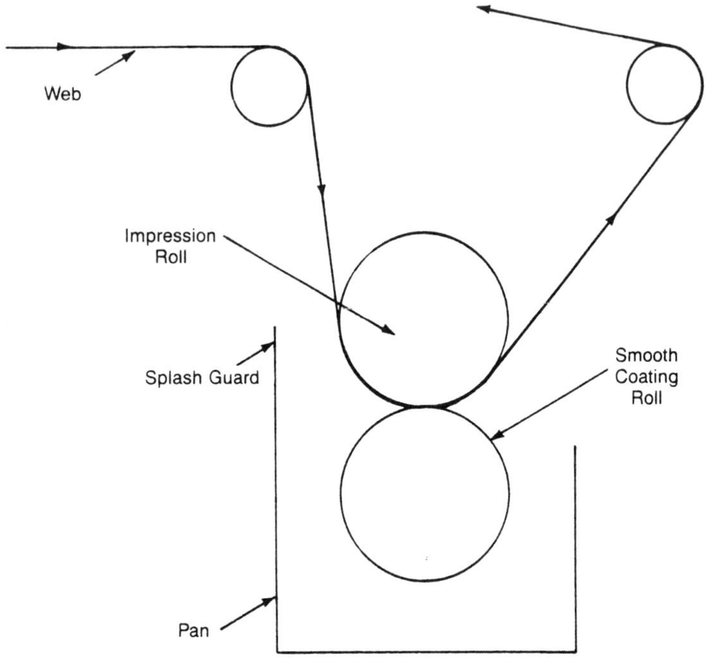

FIGURE 4.7. Direct smooth roll coating.

method. This process is well-suited to running very low viscosity and very low solids coatings, and for running coatings which do not run well by gravure coatings, e.g., water dispersions. The thickness of coating may be determined by using Equation (3.1) in Chapter 3.

In this case, the nip pressure is the pressure exerted by the impression roll. To convert to quantity by weight it is necessary to know the density of the coating. This process may become more important as more and more water dispersions are used.

ROTOGRAVURE

Finally, in Figure 4.8, coating by means of an overall etched rotogravure cylinder is illustrated. It may be used with or without a smoothing bar. These generally run opposite to the motion of the web, are about 3″ in diameter, and are run at about web speed or greater. They make a *kiss contact* with the coated side of the web.

Rotogravure is the most widely used of all coating methods in the converting industry. Engraved cylinders ranging from very coarse (35 lines per inch) to very fine (350 – 500 lines per inch) and three different designs of cell are available, making it possible to apply a wide range of coating thicknesses. It is well suited to apply coatings ranging in dry thickness from 0.0001 – 0.002 inches.

To run well by gravure, a coating should have *length* (i.e., a spatula dipped into it should have a string trailing from it when the spatula is removed, and not droplets), without being pituitous. This feature can also be described as *wet strength*. The coating also must wet the substrate well. These two characteristics will enable the coating to *flow out* when it is removed from the gravure cells. One of the serious problems with water dispersions is that they lack wet strength and do not flow out well without special formulation which may increase their water sensitivity when dried. Good flow out is needed to overcome the tendency for the cell pattern to be obvious to the eye, and to overcome the tendency toward the formation of longitudinal streaks due to drawdown of the coating.

Quantity applied is completely independent of machine speed, web tension, or variations in web thickness but depends only on the depth and number of cells, and the coating formulation. The interfacial tension between the engraved roll and the coating and the web being coated and the coating will determine how the coating will divide between the roll and the web. A doctor blade removes the excess coating, leaving only the engraved cells with coating material.

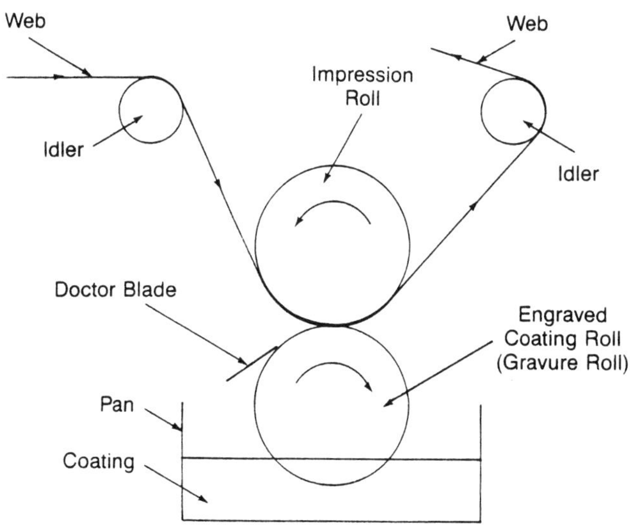

FIGURE 4.8. Rotogravure coating.

Possession of a gravure coater-laminator and a Waldron-Hartig reverse roll coater-laminator will enable the converter to perform almost any coating or laminating job that might come up, if it is one involving the casting of a material that is dispersed or dissolved in the carrier. And if the machines are equipped with drum dryers or tenter-frames, even coating and laminating fabrics and fabric-like materials (non-wovens) would not be impossible.

In running by gravure, it is essential that the rubber impression roll be undercut at the ends so the margins of the web not be coated. This uncoated edge need not be any greater than 1/8″ although it is often as much as 1/4″.

In the cases where dams or deckles are used (as in the case of reverse roll coating), they can be placed in such a way that the edges of the web are not coated.

In all other cases, there is no need to provide for an uncoated edge. If the rolls are exactly parallel to each other, no coating will appear on the opposite side. Also, in a gravure coating head, it is customary to have the doctor blade oscillate back and forth approximately once every 15 revolutions of the roll to insure that no dirt or lumps get trapped under the blade which could result in a longitudinal streak in the coating.

COATING WITHOUT A CARRIER (SOLVENT)—EXTRUSION COATING

Extrusion as a process existed before PE was invented. It had been used in the tire and rubber industry for some time and in insulating wire for electrical applications. When PE arrived, extrusion through a circular die was used for the manufacture of film. Using flat dies for coating a web came later. The process was adopted by some paper makers for multiwalled bags (e.g., St. Regis Paper Co.) and for simple packaging applications (e.g., Thilmany Pulp and Paper Co.). It was also adopted by some converters as another string to their bows. It was, unquestionably, the least expensive way of putting a layer of PE on any substrate.

Extrusion coating equipment consists of two parts: the web handling mechanism comprising an unwind, coating station for application of a primer, drying oven, assorted idlers, rewind stand, and the controls needed to operate such a mechanism properly; and the extruder itself.

The web handling mechanism resembles other web handling mechanisms. What exists on the bulk of extrusion-coating lines resembles a one-color rotogravure press, with the extruder inserted in its proper position.

Screw extruders that function as screw pumps are in universal use in extrusion-coating lines. These convert solid thermoplastic pellets into a viscous melt and force this melt through a die to form a shape. In the case of an extrusion coater, the die is a long, flat orifice, forming the plastic melt into a sheet, which is pressed against a moving web and wound up. The plastic sheet becomes a coating on the substrate.

EXTRUSION-COATING EQUIPMENT

Figure 4.9 shows a typical conventional extrusion-coating line. The substrate is unwound and passed through a coating station where a thin layer of an adhesion promoter, a *primer,* is applied to it. It then enters a drying oven where the primer is dried. The web is then delivered to the nip between a pressure roll and the driven chill roll, where the extruder is located.

In the nip, the sheet of polymer is pressed against the substrate. It is cooled further by making an approximately 270-degree wrap around the chill roll. Then, its edges are trimmed and the finished roll is wound up.

Figure 4.10 shows a section through a typical extruder illustrating the positioning of the screw in the barrel of the machine. Figure 4.11 is a drawing of the screw alone showing its division into the feed section, transition zone, and metering section; Figure 4.12 shows a single flight of the metering section of the screw; and Figure 4.13 shows a cross section of a typical die.

In starting up an extruder, whether for coating or film, it is necessary to heat the barrel, and every extruder has heaters attached to the barrel for this purpose. The applied heat softens the polymer sufficiently to permit the screw to turn.

Since the heaters are thermostatically controlled, they shut off when the temperature reaches the pre-set extrusion temperature. The friction generated by the turning of the screw in the polymer mass is enough to maintain temperature. Once the extruder has started, it is the work of the screw which keeps the polymer molten.

In the extrusion coating of ethylene polymers and copolymers on a substrate, two principal problems are encountered. First, obtaining adhesion of the polymer coating to the substrate and second, maintaining stability of the extrudate, i.e., preventing surging, avoiding starvation of the die, maintaining extrudate temperature, etc.

GETTING ADHESION OF EXTRUDATE TO SUBSTRATE

Even with the use of adhesion promoters, it is necessary to activate the surface of the polymer (PE) sheet exiting from the die to obtain a useful bond between the ethylene polymer (or copolymer) and the web to which it

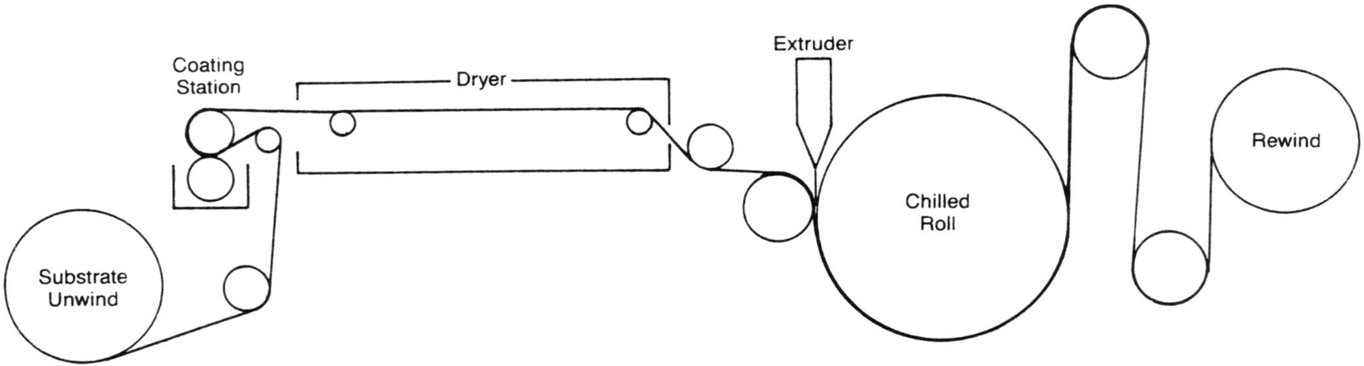

FIGURE 4.9. Extrusion-coating line with priming station.

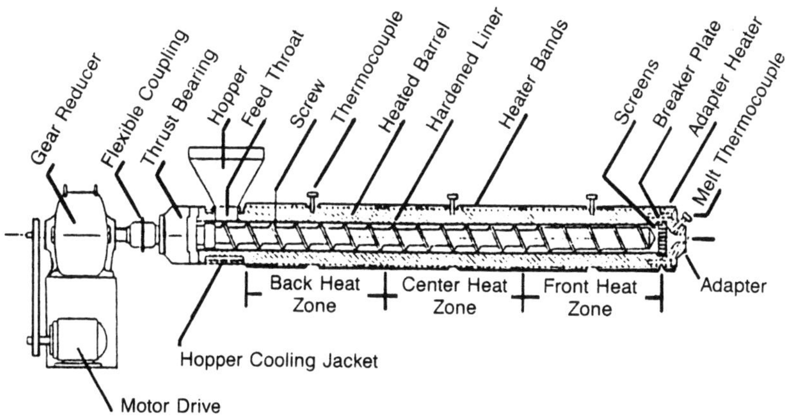

FIGURE 4.10. Diagrammatical presentation of component parts found in all standard single-screw extruders.

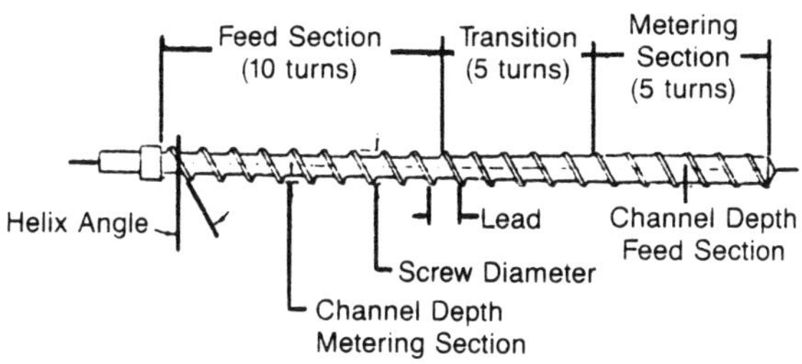

FIGURE 4.11. Metering-type screw. Channel depth decreases to metering section.

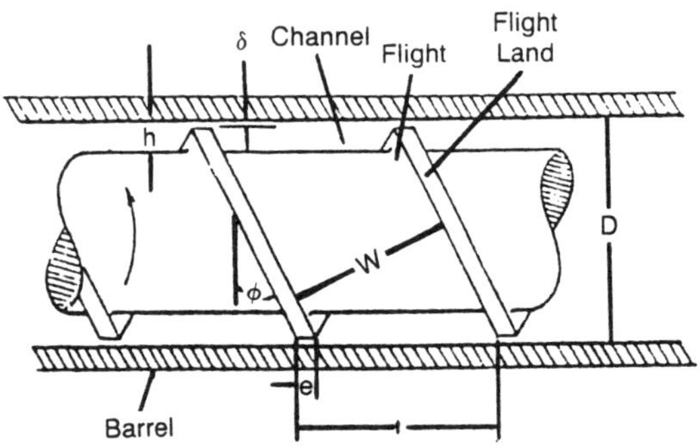

FIGURE 4.12. Points for dimension measurements in extruder screw pump.

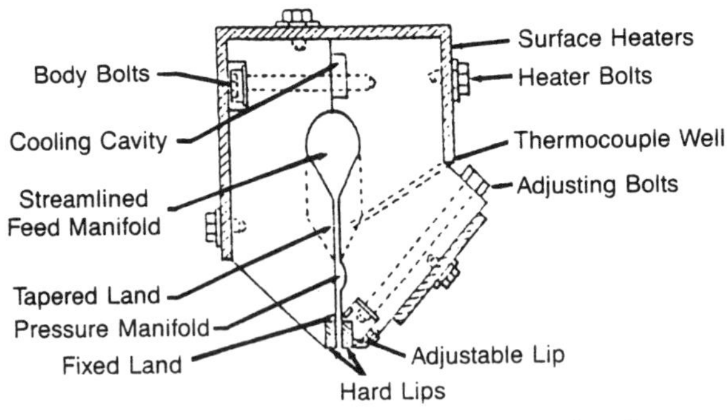

FIGURE 4.13. Flat film and extrusion-coating die.

is applied. This is accomplished by maintaining a high enough temperature in the air gap between the exit from the die and the nip to cause incipient oxidation of the surface of the extrudate.

The mechanism of polyethylene oxidation has been studied and some data have been accumulated. Work on the induction period of the PE oxidation reaction has been done by R. J. Shelton (*ASTM STP 89,* 1949) and B. Baum (*J. App. Pol. Sci.*, 2:281–288, 1959).

The induction period of a chemical reaction is the length of time required, once the conditions for the reaction have been set, for the reaction to begin. It has been found that the logarithm of the time of induction is a linear function of the absolute temperature.

A plot of the data for the induction period of PE oxidation as a function of absolute temperature in degrees Rankin is shown in Figure 4.14. An empirical correlation between log time of induction and the absolute temperature yields:

$$\log t = 16.79210 - 0.01698\,T \tag{4.4}$$

where:

t = induction period (time required for reaction to begin) – sec.
T = absolute temperature – degrees Rankin (460 + degrees Fahrenheit)

The temperature is plotted as degrees Rankin (460 + degrees Fahrenheit). But the numbers inserted are degrees F.

Some calculated values: at 400°F, the induction period is 155 sec., at 500°F it is 3.1 sec. and at 600°F it is 0.062 sec. indicating the high level of sensitivity of incipient PE oxidation to temperature.

To utilize the data, it is necessary to estimate the extrudate temperature in the air gap, where conditions are such, with extrudate width very much larger than its thickness, that the problem becomes a one-dimensional heat flow from the extrudate's surface to its surroundings.

Specifying:

$2w$ = thickness of film (extrudate)
y = distance from die to nip
T_d = temperature of extrudate (film) at die
T_n = temperature of film at nip
T_a = temperature of ambient air some distance away
T_m = mean extrudate temperature
h_c = convective heat transfer coefficient for air

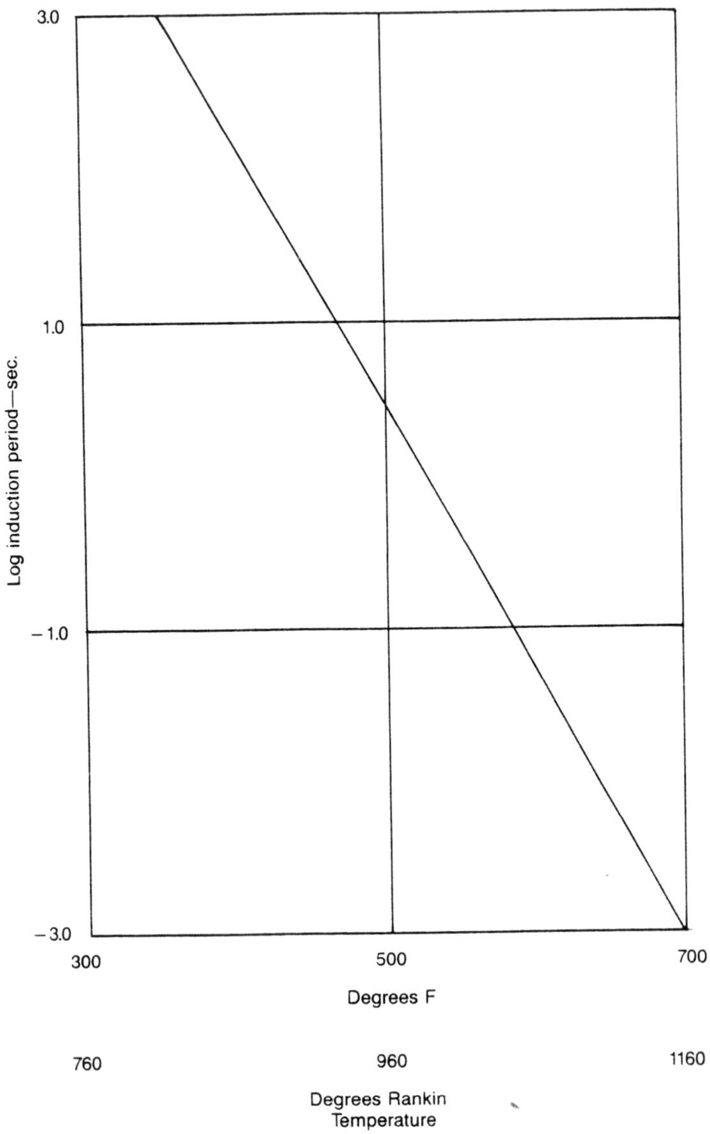

FIGURE 4.14. Log induction period: PE oxidation vs. temperature.

h_r = radiative heat transfer coefficient for extrudate
K = thermal conductivity of extrudate (PE)
k = thermal conductivity of air
$h = h_c + h_r$

A beginning can be made with the general partial differential equation of one-dimensional heat flow:

$$\delta T/\delta t = a\delta^2 T/\delta x^2 \tag{4.5}$$

where:

T = temperature at time t and point x
a = thermal diffusivity of object from which heat is being transferred, in this case PE

The initial and boundary conditions are:

$$T[0,x] = T_d$$

$$T/T[t,w] = -h/k[T(t,w) - T]$$

$$T/T[t,-w] = h/k[T(t,-w) - T]$$

With an extrudate of a given thickness moving through an air gap from a die to a nip, with a given die and a consequent nip temperature, the solution of Equation (4.5) is:

$$[T_n - T_a]/[T_d - T_a] = [4 \sin p_i/2p_i + \sin 2p_i] \exp -p_i^2 s \cos [p_i x/w] \tag{4.6}$$

where:

p_i = a root of the equation $p \tan p = C$
$C = hw/K$
s = dimensionless time variable $= at/w^2$

To make use of Equation (4.6), C must be determined. To find C, it is necessary to know h, the overall heat transfer coefficient expressing the heat lost by the extrudate in the air gap. $h = h_c + h_r$. If h_c and h_r can be computed, it will be possible to proceed with calculation of C, from which p_i may be found.

To demonstrate this exercise, it is advisable to use actual numbers. All units used in making the calculations are in the cgs system. The following are the conditions under which the extruder is to be run:

Machine speed = 450 ft./min. = 228.6 cm/sec.
Air gap = 6″ = 15.24 cm

Then residence time in the air gap is 15.24/228.6 = 0.067 sec. From Equation (4.4), it can be calculated that the mean extrudate temperature must be 595°F. This is the temperature to be used to find h_r.

$$h_r = [T_m^4 - T_a^4]\,\sigma e/[T_m - T_a] \tag{4.7}$$

where:

σ = Stefan-Boltzmann radiation constant = 1.37×10^{-12}
e = emissivity of PE = 0.85
T_m = 595°F = 312.8°C = 585.9 K
T_a = 80°F = 26.7°C = 299.8 K

Then:

$h_r = 4.47 \times 10^{-4}$

Finding h_c is a little more time-consuming, since it must be decided at what temperature to select air properties for determination of two dimensionless numbers—the Reynolds number (*Re*) and the Prandtl number (*Pr*). Since it has been found that under the particular machine conditions, the mean extrudate temperature is 595°F to get incipient surface oxidation and the air temperature has been chosen as 80°F, it seems logical to select the average of the two, or 337.5°F as a reasonable air temperature adjacent to the extrudate and hence the temperature at which to take air properties.

At this temperature:

specific heat = 0.24 = c_p
thermal conductivity = $8.6 \times 10^{-5} = k$
density = $7.8 \times 10^{-4} = \varrho$
viscosity = $2.5 \times 10^{-4} = \mu$

An equation from which h_c may be calculated is:

$$h_c y/k = 0.664\,(Re)^{1/2}\,(Pr)^{1/3} = \text{Nusselt number} \tag{4.8}$$

where:

h_c = convective heat transfer coefficient
y = characteristic length of system; in this case the distance from the extruder die to the nip, or the length of the air gap
k = thermal conductivity of air; material conveying heat from extrudate
Re = Reynolds number = $yV\varrho/\mu$
Pr = Prandtl number, $c_p\mu/k$

Using the numbers given and carrying out the calculation gives:

$$h_c = 1.45 \times 10^{-4}$$

Then:

$$h = h_r + h_c = 4.47 \times 10^{-4} + 1.45 \times 10^{-4} = 5.92 \times 10^{-4}$$

A calculation of radiative heat transfer coefficient and convective heat transfer coefficient at different temperatures shows how strongly the radiative heat transfer coefficient varies with temperature, and how the convective coefficient varies not at all. A plot of these is shown in Figure 4.15. Table 4.2 shows computed values of h_r and h_c at various temperatures.

In Figures 4.16, 4.17, and 4.18 are plotted, respectively: the density, thermal conductivity, and viscosity of air vs. temperature; these properties are used to calculate the two dimensionless numbers.

Some properties of PE at the mean extrudate temperature (595°F) are:

K = thermal conductivity = 8×10^{-4}
c_p = specific heat = 0.659
ϱ = density = 0.757

TABLE 4.2. Values of Radiative and Convective Heat Transfer Coefficients at Temperatures from 100°F to 700°F.

Temperature °F	Radiative Heat Transfer Coefficient, $h_r \times 10^4$	Convective Heat Transfer Coefficient, $h \times 10^4$
100	1.28	1.46
200	1.74	1.47
300	2.25	1.45
400	2.88	1.45
500	3.62	1.40
600	4.52	1.35
700	5.55	1.37

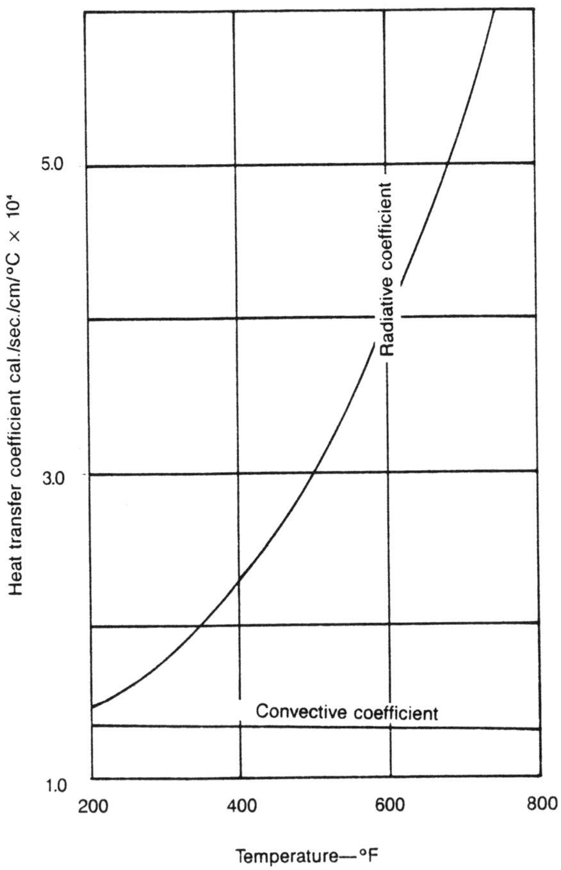

FIGURE 4.15. Radiative and convective heat transfer coefficients vs. temperature.

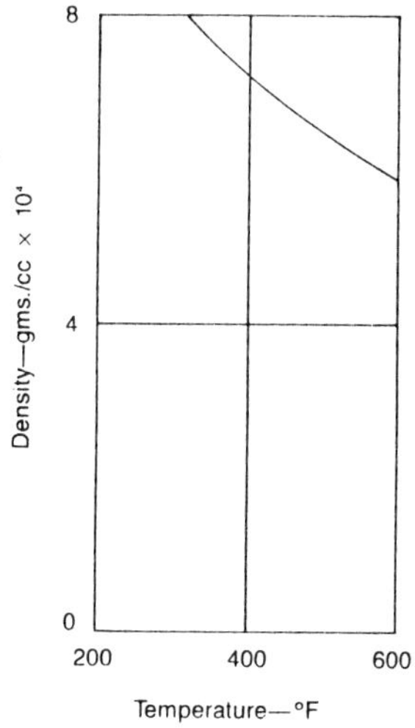

FIGURE 4.16. Density of air vs. temperature.

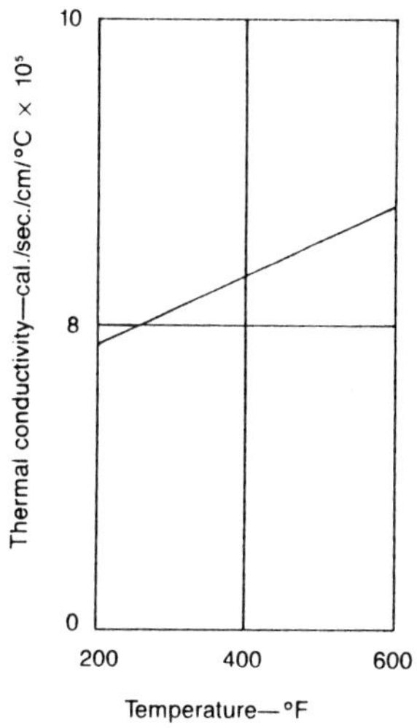

FIGURE 4.17. Thermal conductivity of air vs. temperature.

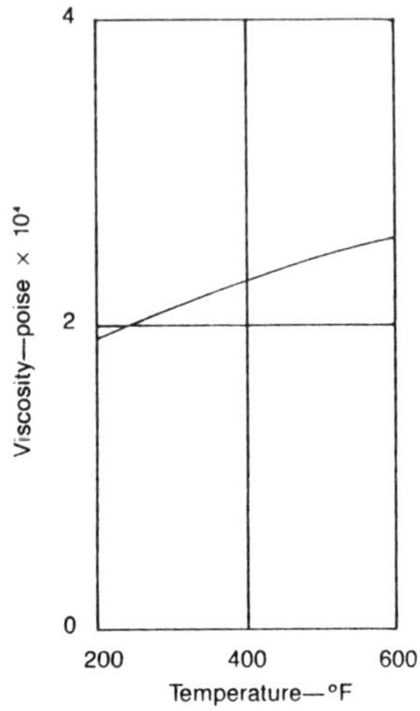

FIGURE 4.18. Viscosity of air vs. temperature.

Then:

a = thermal diffusivity of PE at 595°F $= 8 \times 10^{-4}/(0.659 \times 0.757) = 1.603 \times 10^{-3}$

Assuming the film thickness being run is 0.001 ″ for the particular coating job, then $2w = 0.001'' = 0.00254$ cm. and $w = 0.00127$ cm. Then:

$$C = hw/K = 5.92 \times 10^{-4} \times 1.27 \times 10^{-3}/8 \times 10^{-4} = 0.0009398$$

In *The Conduction of Heat in Solids, 2nd Ed.* (Oxford U. Press, 1959) Carslaw and Jaeger have constructed a table of the first six roots of $p \tan p = C$ for all values of C from zero to infinity. From this table, it is determined that p_1, i.e., the first root of p, is equal to 0.0297, a very small number. The subscript may be dropped, because for small numbers the series in p converges so rapidly that all terms above p_1 may be neglected.

Then $\sin p = p$ and $\cos p\, x/w$ for $x = w$, when w is the surface from which heat is being transferred, is equal to cos 0.0297 = 0.9995589 = approximately 1, and $4 \sin p/(2p + \sin 2p)$ becomes equal to 1. Then, Equation (4.6) becomes:

$$[T_n - T_a]/[T_d - T_a] = \exp -p^2 s \quad (4.9)$$

but

$$p = 0.0297$$

and

$$s = at/w^2 = 1.603 \times 10^{-3} \times 0.067/(1.27 \times 10^{-3})^2 = 66.6$$

then

$$p^2s = 8.821 \times 10^{-4} \times 66.6 = 0.0587$$

and Equation (4.9) becomes:

$$[T_n - T_a]/[T_d - T_a] = \exp[-0.0587] = 0.943 \tag{4.10}$$

T_m may be expressed as the average of T_d and T_n.

$$T_m = [T_n + T_d]/2 = 595 \tag{4.11}$$

Equations (4.10) and (4.11) can be solved simultaneously giving:

$$T_d = 610.1°\text{F}$$

$$T_n = 579.9°\text{F}$$

The indication is that a 600°F die temperature might be too low to get good adhesion of so thin a film. If a die temperature of 600°F were used, the mean extrudate temperature becomes 584.9°F and the residence time in the air gap to get incipient oxidation is 0.112 seconds.

An air gap of about 10″ would give this residence time.

What is often done when adhesion is not quite what is desired is to increase the air gap length, usually with considerable guesswork. While the calculations are somewhat time-consuming, they are quite simple compared to making adjustments on an operating machine.

An alternative is to slow the machine down. In this case, it would be in the ratio 0.067/0.112, i.e., the machine would have to slow to about 270 feet per minute.

Now, let us suppose that the film thickness has been changed to 0.002 inches, keeping all other conditions the same: 6 inch air gap and a speed of 450 feet per minute.

New values of C and p have to be found. Then, solving Equations (4.9) and (4.11) for T_d and T_n gives

$$T_d = 605.5°\text{F}$$

$$T_n = 584.5°\text{F}$$

The basic assumption is made in solving the partial differential equation of heat transfer. Assuming that for small numbers, only one term in the series of roots of the equation $p \tan p = C$ is required permits a solution [Equation (4.9)] of Equation (4.5) to be made rather simply.

This will not introduce a significant error in the calculations.

With the speed of the machine and the air gap length known, and some characteristics of both air and the polymer also known, it becomes possible to calculate what the stock temperature should be to get adhesion of extrudate to substrate. The procedure to be followed is shown below:

(1) From machine speed and selected length of air gap, find residence time in air gap.
(2) From Equation (4.4), find mean extrudate temperature in air gap.
(3) Using this temperature and Equation (4.7), find h_r.
(4) Find h_c from Equation (4.8).
(5) Add h_c to h_r. This is h; the overall heat transfer coefficient for transfer of heat from extrudate to surroundings while traveling through air gap.
(6) Find $C = hw/K$.
(7) From this value of C and any table of roots of p in equation $p \tan p = C$, find the value of p. Since p is very likely to be a very small number because of the nature of the extrusion process, it will be possible to make the various assumptions made and obtain Equation (4.9).
(8) Using Equations (4.9) and (4.11), find T_d, the die temperature needed to obtain T_m, the required mean extrudate temperature.

EXTRUDER OPERATION

The operation of a screw extruder is primarily a function of the last few flights of the screw where the plastic is completely molten, mixed, and of a reasonably uniform temperature throughout its mass. This end is known as the metering section of the screw (see Figure 4.10). The capacity of a machine is defined by the diameter of its barrel, and will vary with polymer (or polymer blend) viscosity at a given horsepower available for turning the screw.

An analysis of a screw pump based on hydrodynamics gives:

$$Q = [F_d D h^2 N^2 (1 - e/t) \sin\phi \cos\phi/2] - [F_p D h^3 (1 - e/t) \sin^2\phi \, P/12\,\mu L] \quad (4.12)$$

where:

Q = volumetric flow rate
D = screw diameter – may use barrel inside diameter
N = rpm of screw (or revolutions per second if more convenient)
h = depth of flight
e = flight thickness
t = screw pitch
ϕ = helix angle (standardized at 17.7 degrees)
P = pressure at discharge end of screw
L = length of metering section
μ = viscosity of polymer at given shear rate

The F's are correction factors which are dependent on values of h/W, where W is the width of the screw flight, and $W = (t - e) \cos \phi$. In the case of modern extruders, these are so close to 1 as to be ignored.

The best way to show how this analysis of a screw pump can be used is to give an example.

Suppose an extruder has the following dimensions:

D = 4.5 inches
h = 0.110 inches
e = 0.375 inches
t = 4.5 inches
$\sin \phi$ = 0.30403
$\cos \phi$ = 0.95266
L = 49.5 inches

Then substituting these numbers into Equation (4.12) yields:

$$Q = 2.919N - 2.68 \times 10^{-6} P/\mu \qquad (4.13)$$

where:

Q is in cubic inches per second.
N is in revolutions per second.
P is in pounds per square inch.
μ is in pounds seconds per square inch.

Since ethylene polymers and copolymers are non-Newtonian in flow characteristics, the viscosity varying with the rate of shear, it is necessary to know the viscosity as a function of shear rate to utilize Equation (4.13).

Figure 4.19 shows apparent viscosity vs. shear rate data for a typical PE polymer. The term *apparent viscosity* refers to the fact that the viscosity plotted is the viscosity the polymer would have, if it were a Newtonian fluid.

The dimensions given are taken from an actual operating machine and the data shown in Figure 4.19 are from a commercially available PE polymer.

Since operation around 600°F is being considered, the data shown in Figure 4.19 are at 600°F.

Shear rate for the screw = *DN/h*.

At any given screw angular velocity, the shear rate is constant and the stock temperature at the discharge end of the die may also be considered constant. Converting *Q* into lbs./hr. and plotting *Q* vs. *P* gives a straight line. This line is called the *screw characteristic* and a different line is

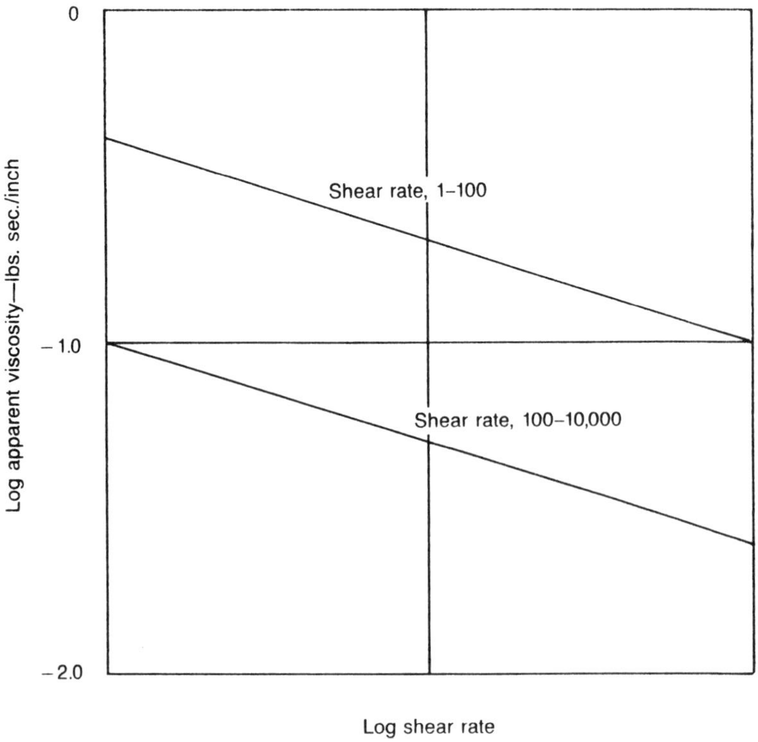

FIGURE 4.19. Log viscosity vs. log shear rate for a typical PE polymer.

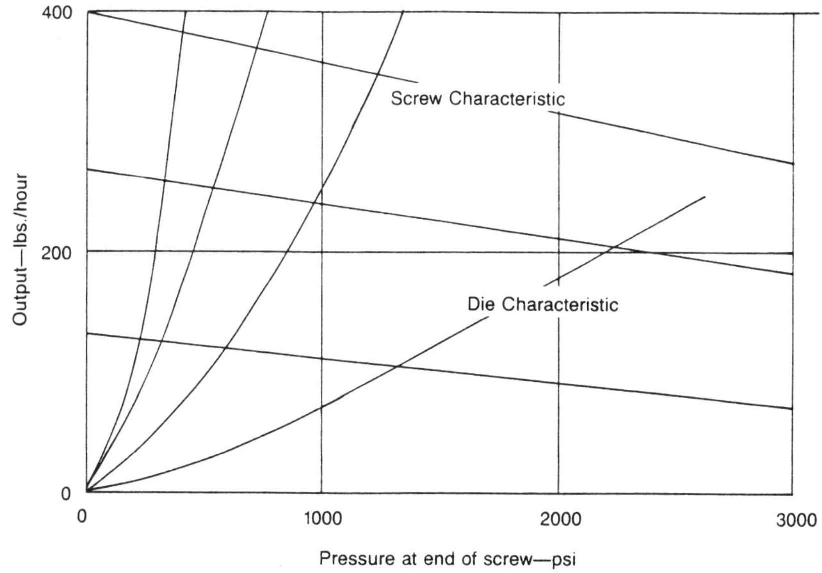

FIGURE 4.20. Screw and die characteristics—output vs. pressure.

obtained for each screw angular velocity. These are shown in Figure 4.20 for four different angular velocities of the screw.

If pressure along the metering section of the screw is measured, it is found that it peaks before the end of the screw. The peak tends to oscillate back and forth with time, producing variations in the quantity of polymer the screw is delivering. This is known as *screw surge* and should be avoided. To do this, it is necessary to provide screen packs and a restrictor valve to create enough back pressure on the screw to cause the peak pressure to be located at its end.

The screw characteristics vary only moderately with increasing back pressure. At 120 rpm, Q, for zero back pressure, is about 540 lbs./hr. At 2,800 lbs./in.^{2}, Q is about 445 lbs./hr.; a change of about 100 lbs./hr. with a change of pressure of 2,800 lbs./in.^{2}, leading to a dQ/dP of about 0.036 lbs. delivered by the screw per lb. per square inch change of pressure. Since the screw characteristics are a set of parallel straight lines, this slope is true of all of them.

Figure 4.21, a cross section of a typical die, shows the dimensions of various parts. These are the actual dimensions of the die used in conjunction with the extruder for which screw characteristics have been computed.

The die consists of four elements:

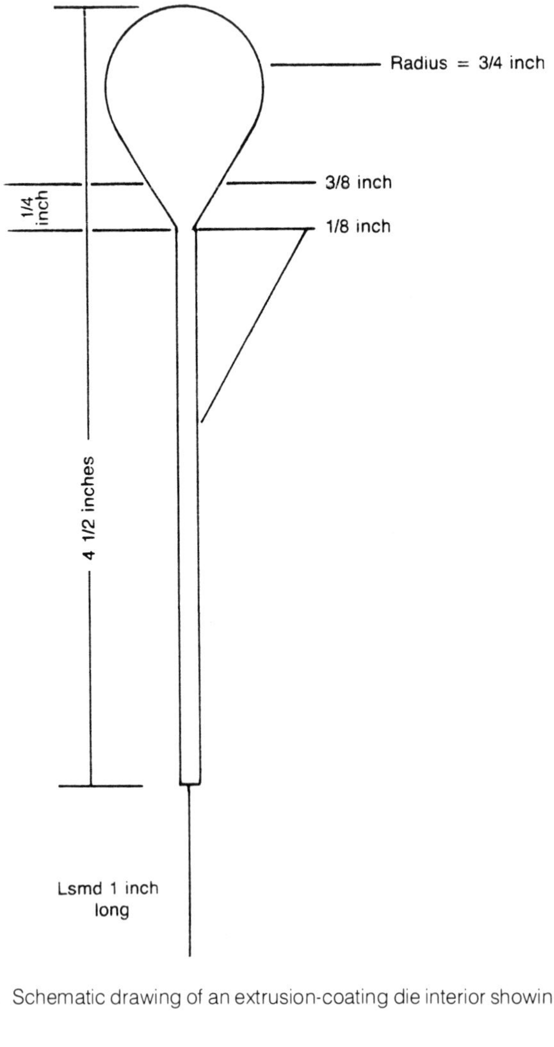

FIGURE 4.21. Schematic drawing of an extrusion-coating die interior showing dimensions.

(1) a 0.75″ pipe
(2) a wedge which may be considered a 0.25″ slot with a 0.25″ opening
(3) a slot 4.5″ long with a 0.125″ opening
(4) a 1″ long land with an adjustable opening

The equation describing flow through a die is:

$$Q = KP/\mu \tag{4.14}$$

where:

Q = quantity pumped through die
K = die constant
P = pressure drop across the die
μ = polymer viscosity

For a pipe:

$$K = R^4/8L$$

For a slot:

$$K = wh^3/12L$$

where:

R = pipe radius
L = length along direction of flow
w = width of slot
h = slot gap

Shear rates (SR) for these configurations are shown below:

For a pipe:

$$SR = 4Q/R^3$$

For a slot:

$$SR = 6Q/wh^2$$

Assuming a 40″ wide web and the die dimensions shown in Figure 4.21, then for the different die parts, the following equations apply:

$$\text{Pipe} \quad Q = 0.006212\, P/\mu \qquad SR = 3.018\, Q \tag{4.15a}$$

$$\text{Wedge} \quad Q = 0.2083\, P/\mu \qquad SR = 2.4\, Q \tag{4.15b}$$

$$\text{Slot} \quad Q = 0.9914\, P/\mu \qquad SR = 9.6\, Q \tag{4.15c}$$

$$\text{Land} \quad h = 0.010'' \qquad Q = 0.00000333\, P/\mu \qquad SR = 1500\, Q \tag{4.15d}$$

$$h = 0.015'' \qquad Q = 0.00001125\, P/\mu \qquad SR = 666.7\, Q \tag{4.15e}$$

$$h = 0.020'' \qquad Q = 0.00002667\, P/\mu \qquad SR = 375\, Q \tag{4.15f}$$

$$h = 0.025'' \qquad Q = 0.00005208\, P/\mu \qquad SR = 240\, Q \tag{4.15g}$$

For each Q, a shear rate can be found. From Figure 4.19, a viscosity for that shear rate can be found, and for that viscosity, a pressure drop may be calculated. If all the pressure drops are added up—pipe, wedge, slot, land—then the total pressure drop across the die for a given output (Q) may be found.

This is dependent on the die opening. The pressure drop through the die vs. output was calculated for four different die openings, and the curve of Q vs. P for each was plotted in Figure 4.20. These curves are known as *die characteristics.*

Figure 4.20 shows screw characteristics and die characteristics. The ideal operating point for an extruder is the intersection between a screw and a die characteristic curve. At that point, screw output is equal to die output and no surging either of the screw or the die will occur.

The relatively high back pressure needed to eliminate screw surge may lead to a die capacity in excess of what the screw is delivering. Thus, at 2,100 psi and 90 rpm, the screw output is about 335 pounds per hour. A die, with an opening of 0.020″ will have an output of 335 pounds per hour at a pressure drop across the die of 660 psi.

Correcting this imbalance requires the installation of a restrictor valve at the end of the screw, the installation of a pressure gauge upstream of the restrictor valve (but not of the screen packs) to read pressure at the end of the screw, and another downstream of the adapter (the part connecting the barrel of the extruder to the die) to read the pressure ahead of the die. The pressure drop through the valve and adapter will serve to balance screw output with die output by adjusting the restrictor valve.

Starting out with an approximately 2,000 psi back pressure and having calculated both screw and die characteristics, it is possible to precalculate running conditions for a given output of the extruder and, by use of the pressure gauges and adjustment of the restrictor valve, to produce a uniform extrudate. Combining these with the calculation of die temperature needed to get adhesion, a very good place to start operation of an extrusion coater is obtained. A recommended back pressure on the screw to prevent surging may be available from the machine's manufacturer.

EXTRUSION COATING AT LOWER DIE TEMPERATURES

Oxidation of PE occurs on both sides of the extrudate as it passes through the air gap. Oxidation of the surface which will be the interior of the package adjacent to the product, i.e., the surface to be heat-sealed, may produce undesirable and undesired effects. The heat seal becomes inconsistent; it can become necessary to change the temperature of the heat-sealing jaws during the course of a run, and more often from one roll of stock to the next. Some off tastes or odors may develop.

In short, it is not wholly desirable to have an oxidized PE next to a food product.

During the early 1960s, when the author was vice president in charge of research and development for Milprint, Inc. (at that time a wholly owned subsidiary of Philip Morris and the largest converter in the U.S.), a project was set up in the Milprint R&D department, under the leadership of Neal Reddeman (now executive vice president of Morton Chemical) and the supervision of the author, to develop a process for extrusion coating in which adhesion of the extrudate, with the use of primers, if necessary, could be accomplished at a temperature low enough so that oxidation did not occur in the air gap. A maximum temperature of 500°F was set as a goal, giving an induction period for oxidation of 3.1 seconds [see Equation (4.4)], far in excess of any practical residence time in the air gap for a normal extrusion coating operation.

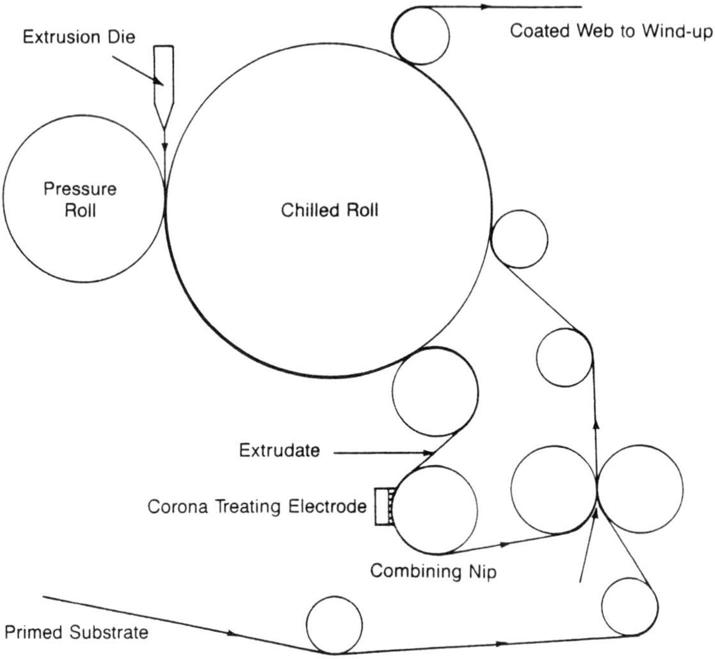

FIGURE 4.22. Diagrammatic sketch of lower temperature extrusion coating process.

The lower temperature, extrusion-coating process was made possible by altering the equipment as shown in Figure 4.22.

If the die land is straight, the extruder may be left where it is but brought in as close to the chill roll as possible; in other words, making the air gap as small as possible. Should the die land be bent, the extruder is located in such a way as to permit the extrudate to fall into the nip. Thus far, the only difference between this process and normal extrusion coating is that the air gap is made as small as possible.

The extrudate is cooled and solidified. It reaches a nip where it has sufficient strength to be peeled off the chill roll. This will occur more rapidly than with conventional extrusion because the die temperature is considerably lower. It is then treated with a CD treater, combined with a primed substrate, primed side to treated side of extrudate, cooled further, the edges trimmed and the coated web wound up.

There are other modifications which may be made depending on the design of the particular machine being altered to suit this process.

The process comprises the following:

(1) Casting a sheet of molten polymer on a chilled roll
(2) Solidifying the melt
(3) Stripping the solidified melt from the chilled roll
(4) Corona discharge treating the stripped film
(5) Combining the stripped, treated film with a primed substrate
(6) Cooling the combined web
(7) Trimming the edges of the web
(8) Winding up the finished roll

All of this is done in a single operation on an extrusion-coating machine.

The process is very well suited to the extrusion coating of polypropylene or other polymers requiring a very short drawdown. Since the extrusion-coating temperature is at least 100°F lower than usual, not as much cooling surface is required to solidify the melt as with conventional extrusion coating.

For Milprint, the use of the process led to a substantial increase in sales of extrusion-coated products. For polymers tending to decompose or depolymerize at higher temperatures, the process offers a way out of many difficulties.

The patent covering the process has long since expired and is available to anyone.

CHAPTER 5

Unit Operation 4 – The Drying of a Moving Web

THE first impression of a newcomer to converting, on entering a converting plant, is the number and sizes of the ovens and dryers by which he is surrounded. The printing presses have ovens built on to them after the last printing station for drying prior to winding up the printed roll. The laminator or coater looks as if it were made up mostly of an oven occupying the entire space between a coating head and a rewind. Even the extrusion coater has a dryer on it following the priming station. (A *primer* is an adhesive applied in a minuscule quantity.) Almost all operations a converter performs, with the obvious exceptions of slitting rolls to size and bag or pouch making, involve passing a web through a tunnel in which some form of heat is applied to it.

And the word *dry* is mentioned quite frequently.

Why?

Because the inks, coatings, adhesives, and primers used by the converters are low-viscosity fluid formulations, usually with solids contents of less than 50%, with some as low as 10–15% (some of the extrusion-coating primers run as low as 1–2%). And all of these, if they are to serve their purposes, must have the fluid, non-solids material, i.e., the solids carrier, removed. In short, they must be dried, whether the carrier is water as in a water dispersion or a mixture of isopropanol and isopropyl acetate, as in the average polyamide flexographic ink.

How dry the web is may determine whether the finished product is acceptable to the customer or not. Retained solvent may result in odor, weak bonds, smearing of inks, corrosion of metal, etc., all of which are undesirable.

(An exception to the need for drying is what is called *hot melt* coating, where a coating is applied from a melt. Coating viscosity is too low for extrusion coating; the coating is therefore applied by equipment specially designed to perform this task. The process is only of passing interest to the converters being discussed here.)

Oven design has a history. Starting as a box with slits on either end through which the web could be drawn with air being blown on it while the web moves through the box, the oven has evolved into today's high-velocity hot air dryers. HVHA is used as an acronym. The evolution has been slow and with some degree of pain, largely because of the general resistance to change that humanity is blessed/cursed with, and also because with each change proposed, oven cost increased. But the rewards were higher speeds for the machines, better quality for the products, and greater customer satisfaction.

A variety of sources of heat have been tried: radiant energy from high-energy electron beams through ultraviolet, infrared, microwave, and radio frequencies; heated plates over which the web is dragged; heated rolls which carry the web with them when rotated, and others even more arcane such as plasma treatment and ultrasonics, and, of course, high-velocity hot air.

All of the drying systems not involving hot air required a hood and blowers to remove the solvent-laden air above the web. Infrared radiators are hot enough to cause a thermoplastic web to melt and a non-thermoplastic one to burn, if the web is stationary. Part of the mechanism needed is a device to swing the radiator away from the web if the machine stops. Some of the other methods suffered from non-uniformity of energy distribution in the web, non-replicability of results, and very expensive equipment. For these and other reasons, HVHA ovens were selected by almost all converters as the drying method of choice.

THE BOUNDARY LAYER THEORY

The theory leading to the use of HVHA dryers was developed early this century. In 1904, a German physicist, Ludwig Prandtl, wrote a paper in which he presented the idea that when a fluid (gas or liquid) flows over a solid surface, there is a boundary layer which adheres to the surface and does not flow with the main fluid stream. When a solid object moves through a fluid, there is a layer at the boundary between the solid and the fluid which adheres to the solid and moves through the bulk fluid with the solid body.

This became important in aerodynamics and airplane design almost immediately. It took some time before it was recognized as being important in any case where a fluid stream passed over a solid surface or where a solid passed through a fluid (stationary or in motion). It did not depend on the nature of the solid or the fluid. If there were relative motion between them, then a boundary layer of fluid, moving with the solid, formed.

The boundary layer theory has found many applications, including the circumstance of a web moving through a hot air stream, as occurs whenever a converter dries a printed or coated web.

The reasoning, as outlined by Dr. Prandtl, stated that frictional forces stemming from the flow of the fluid over the solid surface led to a *no-slip* condition at the surface. As the distance along the surface increased, the thickness of the boundary layer increased as more and more fluid became involved in the flow pattern. Experimental work done on water flowing over thin flat steel plates demonstrated the phenomenon quite well. (A book worth looking at by anyone interested is H. Schlichting's work called *Boundary Layer Theory,* McGraw-Hill, New York, 1955.)

By 1908, enough interest had been stirred up in Prandtl's idea to produce some serious theoretical studies among physicists of the period. Some of this work led to usable formulae for calculation of boundary layer thicknesses. One such, derived from hydrodynamic theory, is that of Blasius for a laminar flow boundary layer. This may be written as

$$z = 5(\nu\, l/V)^{1/2} \qquad (5.1)$$

where:

z = boundary layer thickness (cm)
ν = kinematic viscosity of fluid (cm²/sec.)
l = distance to point at which z is calculated from beginning of boundary layer (cm)
V = velocity of primary fluid stream (cm/sec.)

It should be noted that the units are in cgs. This is not mandatory. Any other system of units will do as well so long as it is consistent. If V is given as ft./sec. then l must be in ft. and ν in ft.^{2}/sec.; z will then be in feet. Since the boundary layer thickness is generally quite small, cm may be the most useful unit. Figure 5.1 illustrates the shape of a curve of boundary layer thickness vs. fluid velocity (in this case, air).

Very interesting, but what does this have to do with drying?

THE DEVELOPMENT OF THE MODERN HVHA DRYER

In the early and middle 1950s, Thomas Gardner, an oven designer, did some extensive experimental work relating air velocity with drying of paper

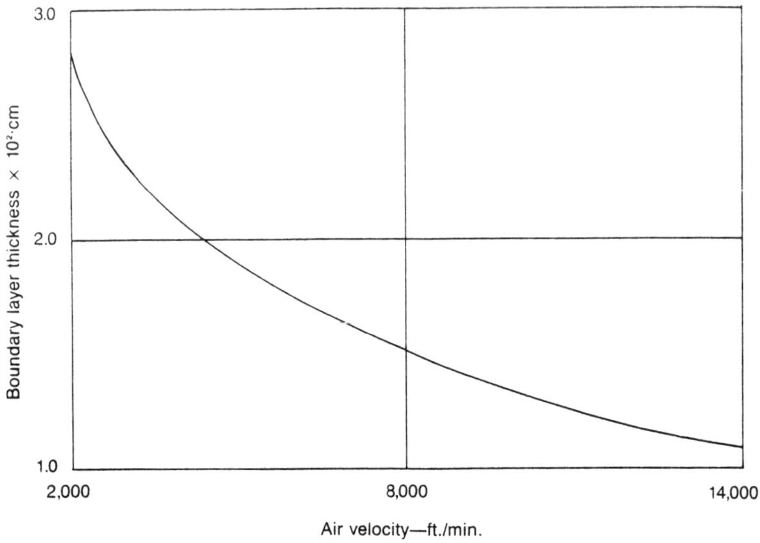

FIGURE 5.1. Boundary layer thickness vs. air velocity.

using air impingement and found that the rate of heat transfer for hot air heating of material moving through an oven follows the pattern shown in Figure 5.2 for the change in rate of heat transfer with change in air velocity (cf. Gardner, T. A. "Performance of Slotted Orifice Air Dryers," *Pulp & Paper Mag. of Canada,* Vol. 62, No. 6, June 1961).

As the air velocity increases, the rate of heat transfer rises, rapidly at first and then more and more slowly as the air velocity rises toward 16,000 ft./min. (readily convertible to cm/sec. when needed). As the air velocity increases, the boundary layer thickness falls, rapidly at first and then more and more slowly as the air velocity rises toward 16,000 ft./min. While the boundary layer thickness approaches zero as a limit, the rate of heat transfer may increase without limit, according to Gardner's work. This is unlikely and it may be that Gardner was unable to reach high enough air velocities to show the curve approaching an upper bound. It appeared as if the rate of heat transfer was inversely related to the boundary layer thickness, i.e., the thinner the boundary layer, the higher the rate of heat transfer; the thicker the boundary layer, the lower the rate of heat transfer. The relationship was not linear.

In any case, it is through this boundary that heat is transported to the web and it is through this layer that the evaporated solvent diffuses to the main air stream and is carried away.

These are two very good reasons for keeping the boundary layer as thin as is feasible.

While at 10,000–12,000 ft./min., a fairly common range of air speeds used in modern HVHA dryers, the main air stream is turbulent, and it is likely that the boundary layer flow regime will be laminar. Air velocity in the boundary layer starts at zero, increases in a laminar fashion, then goes through a small transition zone with the onset of turbulence, and quickly reaches the main air stream velocity and flow condition. It is not too great an approximation to use the Blasius formula to calculate the thickness of the boundary layer. To make it thinner, the air velocity must be increased and/or the kinematic viscosity of the air must be decreased and/or the length of path from air inlet to exhaust port must be made smaller. A clue to some simple principles of oven design and useful air temperature ranges may be obtained from considering methods for minimizing the thickness of the boundary layer.

Figure 5.3 shows the variation of kinematic viscosity (ν) for air with temperature. From the curve, it becomes obvious that raising air temperature increases kinematic viscosity, thus increasing the boundary layer thickness, an undesirable result. But lowering the air temperature reduces the vapor pressure of the solvent, thus reducing its rate of evaporation, a more undesirable result than a slight increase in ν and a consequent increase in boundary layer thickness due to temperature elevation.

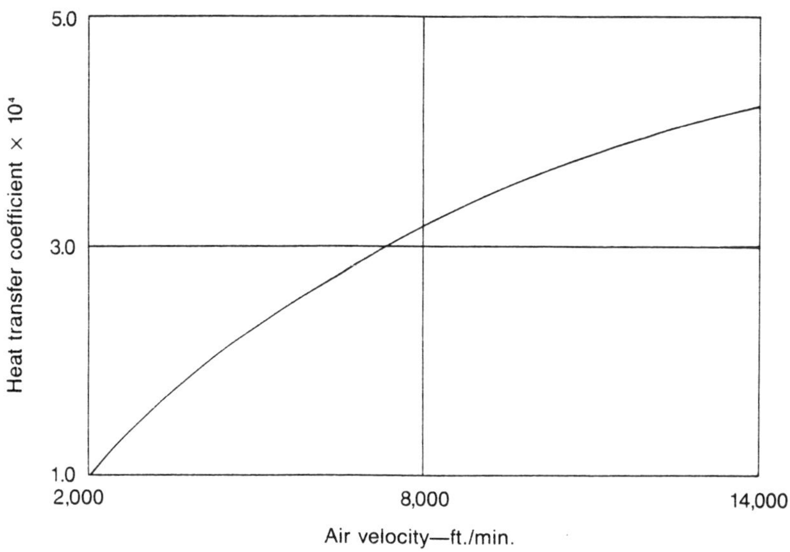

FIGURE 5.2. Heat transfer coefficient vs. air velocity.

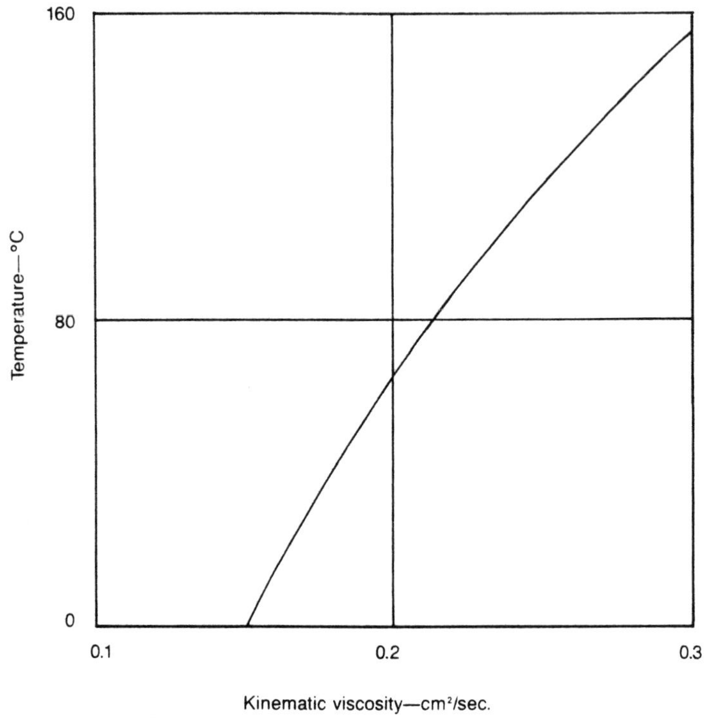

FIGURE 5.3. Kinematic viscosity for air vs. temperature.

Figure 5.4 shows, schematically, how inlet and exhaust ports might be arranged to shorten the length of the air path. In addition to space available, cost considerations enter into deciding how close and how many inlet and exhaust ports will be included in the finished oven.

The critical variables in HVHA oven design are overall length, distance between inlet jet and exhaust port, distance of web from inlet jets, heating capacity of burners, quantity of inlet air, quantity of outlet air, and level of insulation of oven structure. Instrumental requirements are thermometric readings at different points in the oven for the purpose of determining temperature uniformity, pitot tubes for measuring air velocity, and an infrared device for measuring web temperature, or perhaps several such devices so that web temperature can be measured at different points along the travel path of the web. (An infrared thermometric device can read web temperature without making contact with it.)

The distance from inlet to exhaust can range from 1/2 foot to 4 feet. The distance of the web from the inlet air jets can range from 1/2″ to 2″. As the distance between inlet and exhaust is increased, the boundary layer thick-

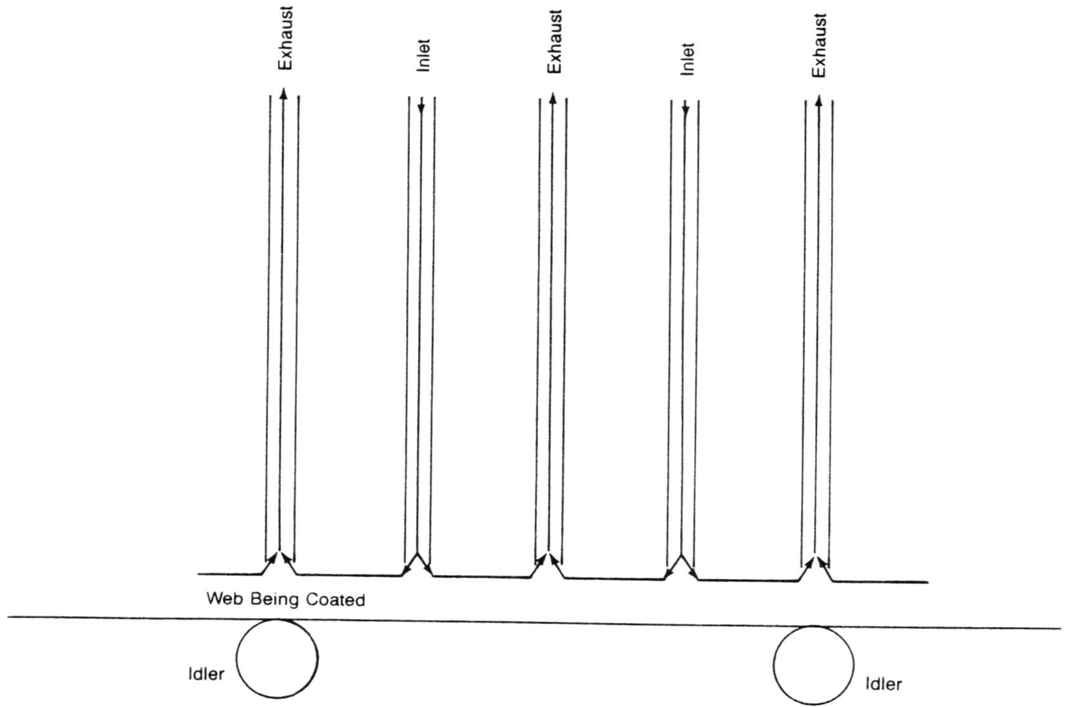

FIGURE 5.4. Schematic design of HVHA oven.

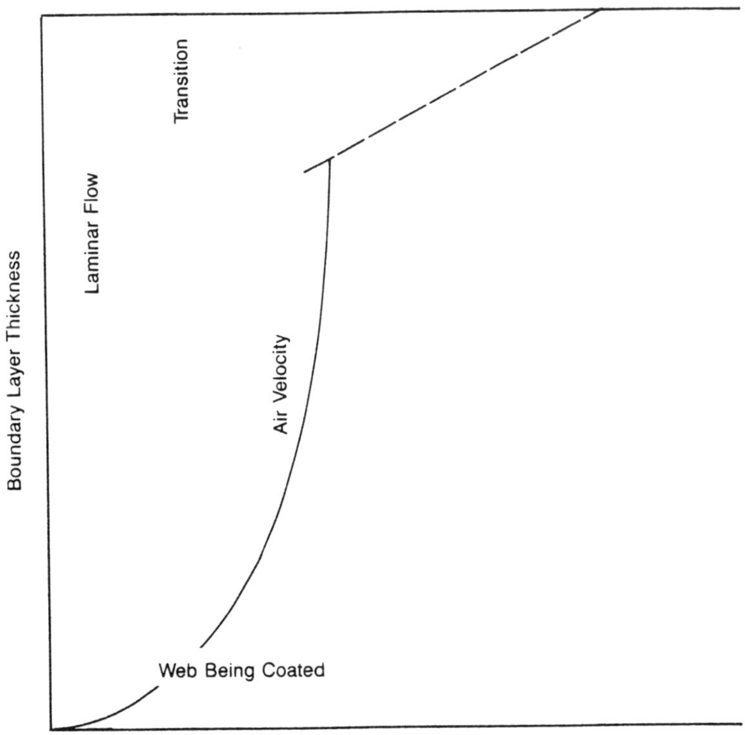

FIGURE 5.5. Velocity profile in boundary layer.

ness increases. As it is reduced, the boundary layer thickness is correspondingly reduced. As air velocity is increased, boundary layer thickness decreases, and as it is decreased, boundary layer thickness increases. Ovens with the smallest distance dimensions and highest air velocities are more efficient than ovens with larger distance dimensions and lower air velocities. They are also more expensive. A satisfactory air velocity that strikes a balance between efficiency and cost is 12,000 ft./min.

As noted above, while the heated air stream might be turbulent at the instant of impingement on the web, within a short time of travel with the web as a boundary layer it becomes laminar and develops a velocity profile such as that shown in Figure 5.5. In all modern ovens, a good deal of the heated air is recirculated to make the operation more economical.

THE QUANTITATIVE ANALYSIS OF DRYING A MOVING WEB

An equation or set of equations enabling the converter to calculate residual solvent and final web temperature would be useful in enabling the converter to correlate initial and final conditions. Even if the equation or equations were not very accurate, they would serve as a starting point for the converter when breaking in a new machine or starting a new job. On the other hand, they might be quite useless because more data about the materials being run are needed than are available, or because some basic phenomenon concerning drying is not sufficiently understood.

Why the converter wants to know the amount of residual solvent is obvious. Final web temperature is important for more subtle reasons. If it is too high, the web may have softened enough to have stretched because of the tension applied in taking it through the oven. This could make for difficulties on the packaging machine on which the web is subsequently run, whether form and fill, or pouch maker, in maintaining cut-off registration of the printed legend on the package. If it is too low, the web might not be sufficiently dry.

Obtaining exact solutions of the heat transfer equations is extremely difficult and has been done in only a few cases. Even the approximate solutions that can be used are difficult to obtain. A great deal of experimental work has been done on heat and mass transfer for many specific cases, and many empirical correlations have been developed for these specific cases. Much of both heat transfer theory and experimental work is covered by Max Jakob in his massive work *Heat Transfer* (John Wiley and Sons, New York, 1949, 1957). Much of both theory and practice on mass transfer is covered in a volume by Bird, Stewart, and Lightfoot entitled *Transport Phenomena* (John Wiley and Sons, New York, 1960). There are many other works of great quality on these subjects by such distinguished writers as McAdams, Eckert and Drake, and others. Unfortunately, almost nothing has been done on the drying of polymer dispersions or solutions when cast on a moving web of material as is done by the converter.

One starting point for an analysis of what is occurring in the oven when a web with a freshly applied wet coating is being drawn through it to be dried is an energy balance, i.e., a non-subtle use of the first law of thermodynamics: the law of the conservation of energy; i.e., heat into the web equals heat stored in it plus heat carried out of it, thus:

<table>
<tr><td>Heat transferred
to web from hot
air stream</td><td>=</td><td>Heat absorbed as
sensible heat by web
and coating components</td><td>+</td><td>Heat consumed
in solvent
evaporation</td></tr>
</table>

This is a simple enough statement. Now it becomes necessary to expand each term sufficiently to give the statement some quantitative meaning.

To do this, English is translated into the symbology and language of mathematics. (It is always slightly surprising how easily the analysis of any physical problem slips into mathematics.)

Standard procedure is to begin with a diagram of the moving web and to focus attention on a small part of it. In this case, the small part shall be taken as 1 cm² (see Figure 5.6). Then, the following may be written:

$$[(k/z) + h_c + h_r]\, T/x = (\text{Sum of } m_i c_i \text{ for elements of the web})\, T/t + Hm_e \tag{5.2}$$

Heat transferred by conduction, radiation, and convection from hot air stream to web	=	Heat absorbed by web and coating components	+	Heat used in evaporation of solvent

where:

- k = thermal conductivity of air at temperature = T_A
- z = boundary layer thickness (average, cm)
- T_A = hot air temperature (Kelvin)
- T = web temperature (Kelvin)
- h_r = radiative heat transfer coefficient
- h_c = convective heat transfer coefficient
- m_w = mass of web (gm/cm²)
- c_w = specific heat of web
- m_p = mass of coating solids (gm/cm²)

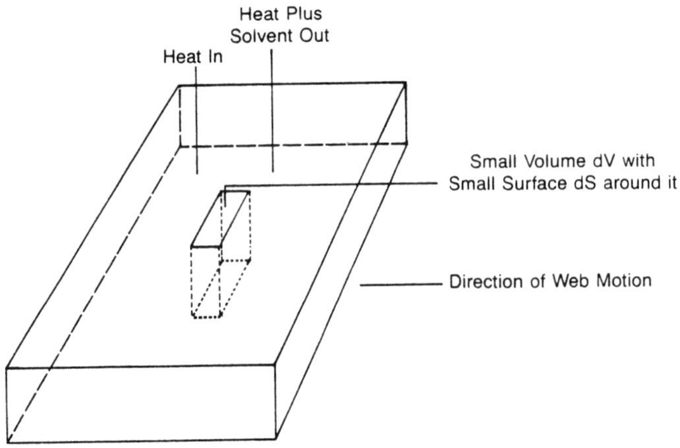

FIGURE 5.6. Beginning of mathematical analysis.

c_p = specific heat of coating solids
m_s = mass of solvent (gm/cm^2)
c_s = specific heat of solvent
m_e = mass of solvent evaporated at temperature = T (gm/cm^2)
H = heat of vaporization of solvent at temperature = T (cal/gm)

If h is equal to $k/z + h_r + h_c$, and MC is set equal to the sum of $m_i c_i$ then Equation (5.2) becomes:

$$h\,\delta T/\delta x = MC\,\delta T/\delta t + Hm_e \tag{5.3}$$

Heat enters the small volume of dV of the web (1 cm on each side and the thickness of the web) across the surfaces surrounding it. Some of the heat entering through the surfaces is stored in this small volume, some of it is used for solvent evaporation.

Then Equation (5.3) becomes:

$$h\,\delta T/\delta x\,dS = MC\,\delta T/\delta t\,dV + Hm_e\,dS \tag{5.4}$$

but by Gauss' theorem FdS = div $F\,dV$ where F is any function:

$$h\,\mathrm{div}\,\delta T/\delta x\,dV = MC\,\delta T/\delta t\,dV + \mathrm{div}\,Hm_e\,dV \tag{5.5}$$

which leads to:

$$h\,\delta^2 T/\delta x^2 = MC\,\delta T/\delta t + H_e\delta m/\delta x \tag{5.6}$$

This is the differential equation that must be solved in order to determine residual solvent and final web temperature.

If terms are rearranged, then the following is obtained:

$$h\,\delta^2 T/\delta x^2 - h_e\delta m/\delta x = MC\,\delta T/\delta t \tag{5.7}$$

This states, mathematically, that heat into the web minus heat lost by it due to solvent evaporation equals heat stored in it.

All units are in cgs. All temperatures are in degrees Kelvin.

In writing Equation (5.7), only a few assumptions, all leading to approximations, have been made, and some complications have been introduced because of three implications. First, it is assumed that the temperature taken transversely across the web along a line perpendicular to the direction of web travel is constant, i.e., that variation in temperature from the left side to the right, along a line all of whose points are equidistant from the oven entrance, does not occur. Second, possible heating of the web from bottom to top is ignored, incorrectly, for at the very least, the oven environment is radiating heat all around the web. Finally, the change in the value of all physical properties with temperature over the range encountered is averaged.

How much error these assumptions will bring about can only be guessed at. Perhaps the error will be minimal, perhaps not. But the error is there, because the assumptions are there.

The equation contains two dependent variables: T and m_e. It becomes essential for the solution of the problem to find how m_e depends on T.

From kinetic theory:

$$r = dm_e/dt = P\,[[MW/2\,\pi\,RT]]^{1/2} \tag{5.8}$$

where:

r = rate of vaporization of a liquid ($gm/cm^2/sec.$)
P = vapor pressure of liquid at temperature T ($dynes/cm^2$)
MW = molecular weight of liquid (gm/mol)
R = ideal gas law constant (ergs/mol degree)
T = temperature (Kelvin)

$dm/dt \cdot dt/dx = dm/dx = dm/dt/(1/V)$ where V = some velocity which is ill-defined. However, we can arrive at the relation:

$$dm/dx = P[[MW/2\,\pi\,RT]]^{1/2} \tag{5.9}$$

and Equation (5.6) becomes:

$$\delta^2 T/\delta x^2 - [MC/h]\,\delta T/\delta t = H/h\,P(MW)/(2\,\pi\,RT)^{1/2} \tag{5.10}$$

The equation still contains two dependent variables. One, the temperature of the web, and the other, the solvent vapor pressure P over the coating. It is conceivable that one might use the classic Clausius-Clapeyron equation, relating vapor pressure to temperature, but that applies only to pure liquids. It is not of any great assistance in these circumstances, since at no time does the problem deal with a pure liquid. The concern is with polymer dispersions or polymer solutions.

The difficulty with the drying problem is that mass and heat are both being removed from the web, as contrasted with extrusion coating, laminating, or heat sealing, where the quantity of mass present is the same before and after the operation is performed.

One of the effects of dissolved or dispersed materials is to lower the vapor pressure of the substance in which solution or dispersion takes place. For solutions of simple substances (ex., sucrose in water), Raoult's law seems to apply, i.e., vapor pressure is inversely proportional to the mole fraction of solute; i.e., the higher the mole fraction of the dissolved substance, the lower the vapor pressure of the solution. And this is true only for dilute solutions.

Deviations from Raoult's law are common in the case of polymer solutions. The polymer dissolved in a solvent, being a molecule of high molecular weight, perhaps with many branches and entanglements, occupies much of the space available to it and the solvent carrier, even when only a small fraction of a mole is present. It occupies a fair amount of the surface of the solution and reduces the area available for solvent evaporation. As the coating is dried, more entanglements result from the scrambling of polymer segments. The segments lose more and more of their mobility and begin to produce random sizes and shapes of very small spaces in which molecules of solvent or clusters of solvent molecules are trapped.

As drying proceeds further, these tend to become still smaller, trapping solvent ever more tightly. Experience teaches that drying becomes more and more difficult as the solvent concentration is reduced and that removing the last few percentages is the most difficult. This is the fundamental reason for the acceptance by the market, the industry, and even the FDA, of an amount of residual solvent greater than zero. This generally is given as mg per ream. A ream is 500 sheets 24″ × 36″ or 432,000 in^2, a unit borrowed from the paper industry.

An acceptable level is 300 mg/ream. At this level, the solvent is so tightly trapped within the dried coating that the deleterious effects of retained solvent are absent.

Some attempts have been made to put the reduction of vapor pressure of polymer solutions on a quantitative basis. Flory has developed a theory built around the use of volume fraction in place of mole fraction (cf. Flory, *Principles of Polymer Chemistry*, Cornell University Press, 1953). Huggins and Flory each independently developed the idea of a polymer-solvent interaction coefficient, but these are difficult to determine and data exists for only a few solvent-polymer pairs.

However, all of these corrections do not account for the large reduction in vapor pressure encountered in the actual drying of a coated substrate.

Perhaps a closer examination should be made of what is happening when a freshly coated substrate is being drawn through an oven where the coating will be dried. While the substrate is being heated all over – top, bottom, and ever so slightly, at the edges, due to radiation from the hot air and the oven walls – the primary heating is being done by the hot air impinging directly on the coating surface. Evaporation from the surface occurs more rapidly than fresh solvent rises to the surface, perhaps because of the slight lowering of temperature from the top surface of the coating downward to the bottom. The consequence of this temperature gradient from the bottom to the top surface of the coating is that the surface concentration of solvent (or dispersant) becomes lower than that of the wetter coating below it. There exists a concentration gradient of solvent from the bottom of the coating to its top.

As drying proceeds, the difference in concentration of solvent at the surface and in the interior of the coating tends to increase. Because of solvent loss, the surface forms a skin of polymer, some small quantity of solvent, not yet evaporated, plus whatever other components are present in the formulation which are non-volatile. The solvent below this membrane must now be driven through it in order to evaporate. The rate of drying is drastically reduced, since the solvent below the surface skin must all but diffuse through it in order to evaporate, a process much slower than vaporization even if the rate of diffusion is quite high for a diffusion process. It should be noted that the word solvent refers to both solvents and dispersion media, e.g., water.

Every converter has had experience with this skinning tendency. Sometimes the substrate will be taken through the oven a second time in an effort to drive off more solvent. Sometimes the coating will be heated above its softening point to insure dryness. This can be done only with heat resistant substrates.

The tendency to skin varies from one polymer-solvent combination to another, and from one water-dispersion to another. It is also dependent on other substances in the formulation. But virtually all the formulations used by the converter show this tendency. And there is, as yet, no way to predict, quantitatively, either its extent or its effect.

If a relation is found between vapor pressure and temperature for a polymer solution or dispersion through some other characteristic of a solution or dispersion, e.g., viscosity change with concentration, then vapor pressure change with viscosity and viscosity change with temperature, then by going a long way around, vapor pressure and temperature, for what is of concern here, will be functionally related to each other.

It is known that the evaporating solvent concentration will change from its initial values to very close to zero when the web exits from the tunnel. Then, by making all these substitutions in Equation (5.10), assuming that Equation (5.10) can be solved, a relationship accounting for skinning, the resulting reduction in vapor pressure and the consequent reduction in drying rate is found. To do something of this sort would require a good deal of experimental work to accumulate data. There is no indication that this work is being done.

Starting from rather simple mathematics, the problem of developing an equation or set of equations describing the drying of a freshly coated web by drawing it through a reasonably efficient HVHA oven is anything but simple.

But the converter can help himself. What can be done is to instrument the equipment and record the data on each run he makes: what was run, air temperature and air velocity, initial web temperature, machine speed, final web temperature and residual solvent. The supplier will tell the converter about solids content and viscosity of the coatings and adhesives dealt with; or these can be readily determined by the converter.

To do this properly, the converter will need to measure air velocity (pitot tube), machine speed (something the converter does anyway), air temperature (another measurement the converter makes in any case), web temperature (an IR thermometer), and residual solvent (something which can be done very nicely with a temperature programmed gas chromatograph).

From the accumulated data, the converter will be able to develop correlations between initial wet web conditions and final dried web conditions; specifically, the final web temperature and residual solvent for the materials run and the machines they are run on.

Should the converter have a quality assurance lab, which every converter ought to have, all of this can be done at minimal expense. The work will pay off in time saved setting up a machine and improved product quality. It will also tell the converter a good deal about the limitations of the equipment used.

CHAPTER 6

Unit Operation 5 – Laminating

MANY products require special characteristics, such as a very strong seal, a low water-vapor and equally low oxygen permeability, perhaps some additional strength and stiffness, etc. It is all but impossible to achieve many of these needs with any one of the materials available to the converter today.

So the converter is often called on to perform the act of combining several substrates together, as in the package for Pepperidge Farm™ cookies. More often it is a two-ply combination, as in a typical snack food package. The process for doing this combining is called laminating.

An obvious example is paper-backed metal foil.

The strata selected serve many purposes. Their properties are so very different.

LAMINATING FOILS—METAL AND PLASTIC

Making a 0.00035″ (or thinner) aluminum foil useful demands that it be supported to prevent it from wrinkling and tearing when subjected to even the mildest handling. Without the support, the metal foil is nice to look at but serves no useful packaging purpose.

PE is printed and then laminated to itself to form a rice bag, with printing buried between the two layers and the adhesive acting as a barrier to the migration of rice oil, which PE resists only slightly. A lamination of two 0.0015″ films serves the application well. A single 0.003″ web fails miserably.

While processes for backing metal foils with paper (paper-backed tin and lead-tin composition foil: a core of lead and surfaces of tin) have been in existence since late in the last century, laminating has become more extensively used since the end of World War II and the ever-increasing applications of polyolefins.

And again, while paper-backed metal foil was begun using the tin and modified tin foils mentioned above, it has extended itself to aluminum foil as aluminum pushed out tin. Originally, aluminum foil was laminated either with a wet adhesive (a high concentration of glucose in water was one of the very early adhesives) or wax. In recent years wax has fallen into disfavor. To a large extent it has been replaced by extrusion-coated PE, acting as an adhesive to bond paper to aluminum foil for applications such as packaging dehydrated soups and pet foods.

Paper-backing of metal foil is done largely either by wet laminating or the already mentioned use of PE as an adhesive. Wet-laminating gets its name from the fact that the two strata are combined without drying the adhesive, but by evaporating off the water through the porous substrate with which the metal foil is combined.

The use of the process has been extended to the lamination of any impervious web (e.g., PE, nylon, PET, etc.) to any porous web (non-wovens, paper).

LAMINATED PRODUCTS

The use of laminated products for packaging has become quite widespread, perhaps beginning with pharmaceuticals and extending to vacuum-packed meats, where one layer acts as the gas barrier (PVdC-coated cellophane, polyester film, nylon film, PVdC coated, sometimes uncoated) and the other acts as a strengthening and heat-sealing web (PE of thickness ranging from 0.0015″ to 0.0025″, depending on the weight of the meat to be packaged).

In a sense, when a web is extrusion coated with a polyolefin polymer, a product almost equivalent to a lamination is formed, with the caveat that the contribution to properties of the adhesive layer is not present, unless the priming station is capable of coating and drying one to two pounds per ream of primer. This is not generally the case. There are some laminations having properties quite superior to those of extrusion coating. But usually, the two types of product compete for the same market.

LAMINATING—WET AND DRY

There are two kinds of laminating processes: wet and dry. Wet laminating, already described, is illustrated in Figure 6.1. It originally served to produce

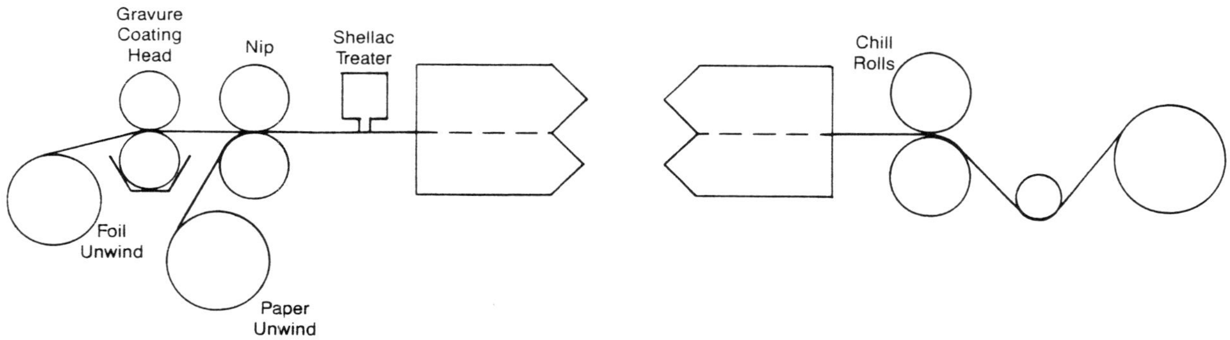

FIGURE 6.1. Wet laminating assembly (schematic).

a paper-backed metal foil for candies and tobacco products. In this process, an adhesive is applied to the metal foil (often by gravure) or other non-porous web, and combined with the paper or other porous web. The combined web is drawn through a drying tunnel and the solvent is dried through the paper (or other porous web). The adhesives used are water-borne, so there is no risk of residual solvent odors. A very popular adhesive is neoprene latex thickened with casein, available from many adhesive suppliers.

Wet laminating can be used for combining almost any film to paper, if an adhesive dispersed in water can be found.

Referring to Figure 6.1, it becomes obvious how the process works. The metal foil or polymer film is passed through the gravure-coating head and an adhesive applied to its underside. The paper, or other porous web, is brought up beneath the foil and combination takes place in a nip, which may or may not be heated. The web is then drawn through the drying oven, is cooled by going through a pair of cooling rolls, and is wound up as a finished roll of stock, ready for the slitter, or whatever the next operation might be.

The adhesive must not be too low in viscosity since it might then penetrate the paper too rapidly. Yet, if the viscosity is too high, it might coat poorly. A satisfactory viscosity is in the neighborhood of 100 cps. On most wet-laminating machinery, there is sufficient open time to permit some of the water to evaporate prior to combining the two webs and allowing for a strong enough wet bond to form to prevent delamination on entering the oven. Usually this process does not work with coated, supercalendered stocks because of their resistance to the passage of water through them. The process has some serious limitations.

In addition, if the foil side of the combination is to be processed further (printed or coated), immediately prior to being taken through the tunnel, a solution of shellac or other adherent material is applied with a wick to the foil surface to protect it from contact with the paper. It has been found that if this is not done, getting adhesion of inks or coatings to the foil is extremely difficult. Whatever is used should be compatible with the inks, coatings, or adhesives to be used later.

THE DRY BONDING PROCESS

The dry bonding process is the major method used by converters to combine two (or more) webs together. The primary web is defined as the one to which the adhesive is applied. The secondary web is the one bonded to the adhesive on the primary web. The adhesive is applied to the primary web by one of the methods described in the chapter on coating and dried as described in the chapter on drying (Chapters 4 and 5, respectively).

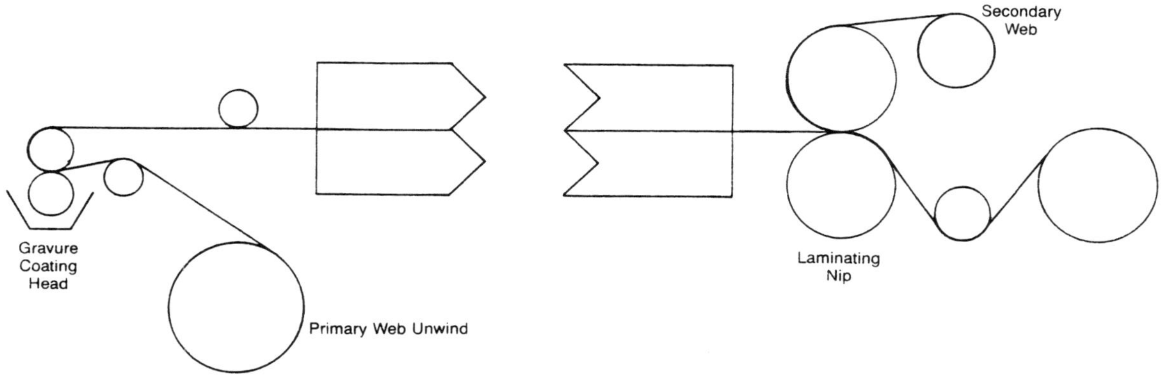

FIGURE 6.2. Dry bonding process (schematic).

The primary web leaves the oven and is combined between a pair of nip rolls, one of which is heated. One roll is bare steel, usually the heated one; the other is steel covered with about a half inch of an elastomeric composition (neoprene, nitrile rubber, et al.) formulated to be heat resistant and cured to a Shore Durometer reading of 85–90–in other words, quite hard, but compressible. The process is illustrated, schematically, in Figure 6.2.

The laminating assembly is generally located forward of the oven exit, leaving enough space to accommodate the mechanism, including the drive needed to run one of the pair of rolls while the other idles. This prevents wrinkling of the secondary web due to some extremely slight difference in velocity of the two rolls, if both were driven. In all nips on converting equipment, one roll is driven, while the other idles.

The rolls need not be of the same diameter (since one idles) although they often are. The laminating unit also serves as a pull unit for the coater, since it draws the primary web through the oven and maintains tension on it. The drive of the laminating assembly is provided with a mechanism that enables it to speed up or slow down, adjusting the web tension to a preset value.

There are drives available which can do this automatically by having each tension point (pull unit feeding the web to the coating head, coating head, laminating assembly, and rewind) driven by separate direct current (DC) electric motors and synchronizing the motion of all of them. It is designed so that the motors talk to each other and adjust their speeds from the signals they receive. Such drives, all DC, have been made commercially by Reliance Electric Co., the Louis Allis Corp., and others, for some time.

Because of the distance of the laminating assembly from the oven exit, the primary web tends to cool considerably while traveling from one point to the other. To maintain temperature, it is often wrapped around the heated roll as much as is allowed by the geometry of the machine.

ALTERNATIVE CONFIGURATIONS OF LAMINATING ASSEMBLIES

There are many ways to arrange rolls to do *dry bonding,* as this method of laminating is sometimes called. Figure 6.2 illustrates one. It is a sketch of a complete piece of equipment showing a coating head, an oven, a laminating assembly and a wind-up. Figures 6.3 and 6.4 show only the laminating assembly and rewind for two other possible configurations.

If it is considered that the web is a plane plate and that the surface of the roll is projected onto a plane adjacent to the web after it leaves the tunnel, as shown in Figures 6.3 and 6.4, then the matter of heating the plate becomes a unidirectional flow of heat from the roll surface into the web of material with the adhesive as the outer layer. Figure 6.3 shows an arrangement

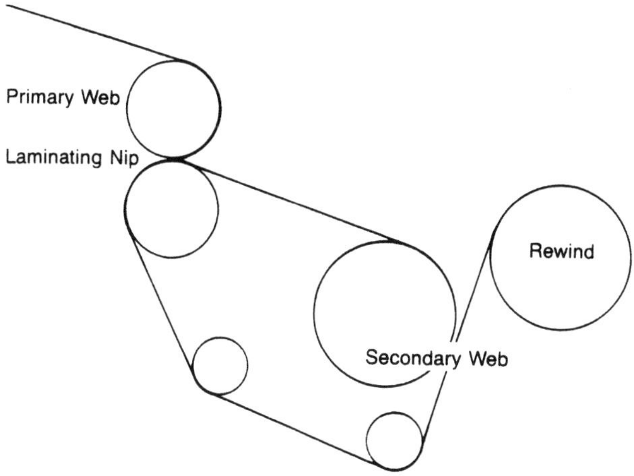

FIGURE 6.3. One roll configuration.

whereby the web may be wrapped around the heated roll to a matter of around 135 degrees or less, while Figure 6.4 permits of no wrap. Should the adhesive require both heat and pressure to bond to the secondary web, wrapping the heated roll becomes a vital part of the laminating process.

AN ANALYSIS OF HOW HEAT IS TRANSFERRED FROM HEATED ROLL TO WEB

The equation, which has been seen before in this work, governing the transfer of heat with one roll heated and the other neither heated nor cooled but allowed to adjust to ambient conditions is:

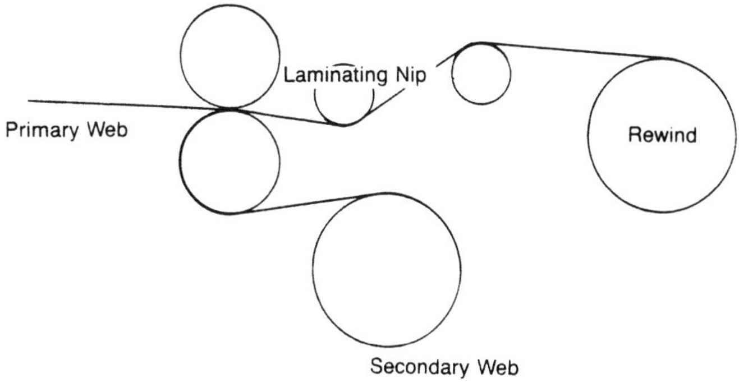

FIGURE 6.4. Another roll configuration, no wrap.

$$\delta T/\delta t = a\ \delta^2 T/\delta x^2 \tag{6.1}$$

where:

x = the coordinate along which heat is flowing
T = the temperature at x, t
t = time
a = the thermal diffusivity of the material being considered. It equals $k/\varrho c$, where k is the thermal conductivity, c the specific heat and ϱ the density of the material. Its dimensions are cm^2/sec. in cgs.

While the conclusion that the greater the wrap, the higher the temperature at the nip where the two webs are joined, is obvious, a solution of Equation (6.1), no matter how approximate, will give some quantitative sense of the significance, or lack of same, of pre-heating the primary web.

A solution to Equation (6.1) in the case of an isothermal boundary layer, which is the case when we consider a thin web in direct contact with a heated metal roll, may be written as:

$$[T_r - T]/[T_r - T_a] = \mathrm{erf}\ x/2(at)^{1/2} \tag{6.2}$$

where:

T_r = temperature of heated roll
T = temperature of web at bonding surface after period of contact with heated roll
T_a = ambient air temperature surrounding laminating assembly
w = thickness of web plus adhesive layer
t = time of contact of web with heated roll
a = thermal diffusivity of composite, i.e., web plus adhesive layer
x = distance into web from zero, which is point of contact of web with roll surface
erf = designation for the error function which has been calculated and whose values are listed in tables, one of which is published by the National Bureau of Standards under the title *Tables of Mathematical Functions,* publication #55 of the Applied Mathematics Section

How Equation (6.2) may be used can best be illustrated by taking an example.

Suppose it is desired to laminate 0.0005 PET (polyethylene terephthalate) to 0.002″ PE with one lb./ream of an acrylic adhesive. In order to activate the adhesive a minimum temperature of about 130–135°F is required at the point of lamination. Should the web wrap the roll, and if so, to what extent? Assume a 10″ diameter roll (radius = 5″).

Suppose the calculation is made obversely. Given a series of different angles of wrap, what will the different temperatures be? If this is determined, then a curve of angle of wrap vs. adhesive temperature at the nip may be plotted and the temperature easily read.

- Find properties of composite, i.e., PET plus adhesive layer. This includes thickness, specific heat, thermal conductivity, density. The thermal diffusivity may then be computed.
- These properties may be calculated as follows: multiply PET property by thickness of PET. Add to this the particular adhesive property multiplied by thickness of adhesive. Divide sum by total thickness–PET plus adhesive. Thickness of adhesive may be found by dividing weight per ream (in this case 1 lb.) by the adhesive density obtaining volume of a ream of adhesive. Dividing this by 432,000, the area of one ream, yields adhesive thickness.
- Find time of contact of web with roll. This is equal to angle of wrap divided by 360, giving fraction of circumference covered by primary web. Multiply this by circumference of roll in feet. This gives distance traveled around roll in feet. Divide this by the web speed in feet per second and the result is the time of contact in seconds.
- Calculate a. This is $k/\varrho c$ for the composite.
- Find $w/2(at)^{1/2}$. Surface where $x = w$ is one where bonding to secondary web is taking place. Look up error function in tables. Interpolate linearly to obtain more exact values. This gives value of ratio of temperature differences.
- Calculate value of desired temperature.

Units used for thermal conductivity, density, specific heat, and thickness are in cgs.

The assumptions made in following this procedure are:

- After coming out of the tunnel, the web cools to ambient air temperature because of the distance the web travels in order to accommodate all the mechanism required for operating the laminating assembly and wind-up.
- It is considered mathematically and physically correct to view the web as an infinitely long flat plate, enabling use of Cartesian coordinates in place of cylindrical ones, thus making calculations simpler.

- The flat spot formed in the nip is assumed to be 0.2 inches and in making the calculations is included as part of the wrap. In other words, there is no zero wrap. The minimum is 0.2 inches.
- Considering heat as transferring through a composite whose properties are calculated as described above instead of considering heat as transferring in individual steps (first through one layer of the composite, and then through the second layer) does not introduce a serious error.

Using the procedure shown above, a table of calculations based on Equation (6.2) can be constructed for angle of wrap vs. temperature of exposed adhesive surface with some of the intermediate calculations. Temperature of the heated roll is taken as 160°F. Those calculations are shown in Table 6.1.

Figure 6.5 is a curve of temperature of exposed adhesive surface vs. angle of wrap. It can readily be seen that a marked reduction in temperature occurs when the wrap angle is quite small (10° or less), and that a substantially higher temperature is reached with some wrap than with none at all.

Now, let us suppose that space requirements are such that the configuration shown in Figure 6.4 is mandatory and that no wrap around the heated roll is possible, or that the coated web cannot tolerate being wrapped around the heated roll without wrinkling. In order to get the most out of the adhesive, the lamination must be made at a minimum of 130°F. At what temperature must the heated roll be maintained?

Since the duration of contact has already been computed for zero wrap (nip only) and was found to be 0.00222 seconds, the function $Z = w/2(at)^{1/2}$ is found to be equal to 0.530 and erf Z = 0.540. Then the ratio $[T_r - T_w]/$

TABLE 6.1. Values of Temperature of Exposed Adhesive Surface as a Function of Angle of Wrap of Coated Web around Heated Roll.

Roll Temperature Taken as 160°F			
Angle of Wrap (degrees)	Length of Contact Including Flat Spot (feet)	Duration of Contact (sec.)	Temperature of Exposed Adhesive Surface (°F)
135	0.998	0.133	153.8
90	0.691	0.0895	152.5
60	0.453	0.0604	151.0
30	0.235	0.0313	146.6
10	0.0897	0.012	139.8
0	0.0167	0.00222	116.3

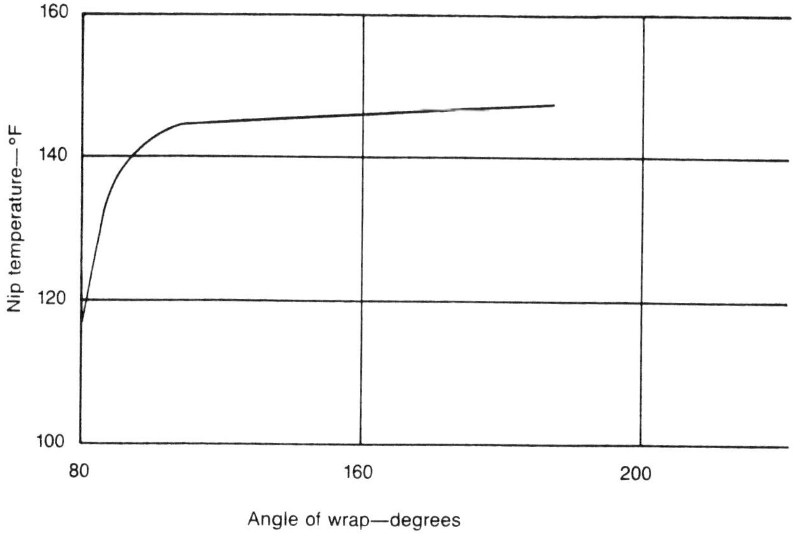

FIGURE 6.5. Nip temperature vs. angle of wrap. Heated roll at 160°F.

$[T_r - T_a] = 0.540$ and T_r may be found with T_w set equal to 130°F and T_a held at 80°F. It is found to be 188.7°F, considerably higher than that required when a wrapped roll is permissible.

Continuing the process of calculation, it is possible to find the whole set of roll temperatures required to provide a given nip temperature. The same thing may be done for any given wrap angle.

If only the nip is considered, then Table 6.2 can be constructed of the roll temperatures required to sustain a given nip temperature.

A graph of Table 6.2 is shown in Figure 6.6, demonstrating how the

TABLE 6.2. Roll Temperature Required to Sustain a Given Nip Temperature at Exposed Adhesive Surface.

Roll Temperature (°F)	Nip Temperature (°F)
123.5	100.0
145.2	110.0
167.0	120.0
188.7	130.0
210.4	140.0
232.2	150.0

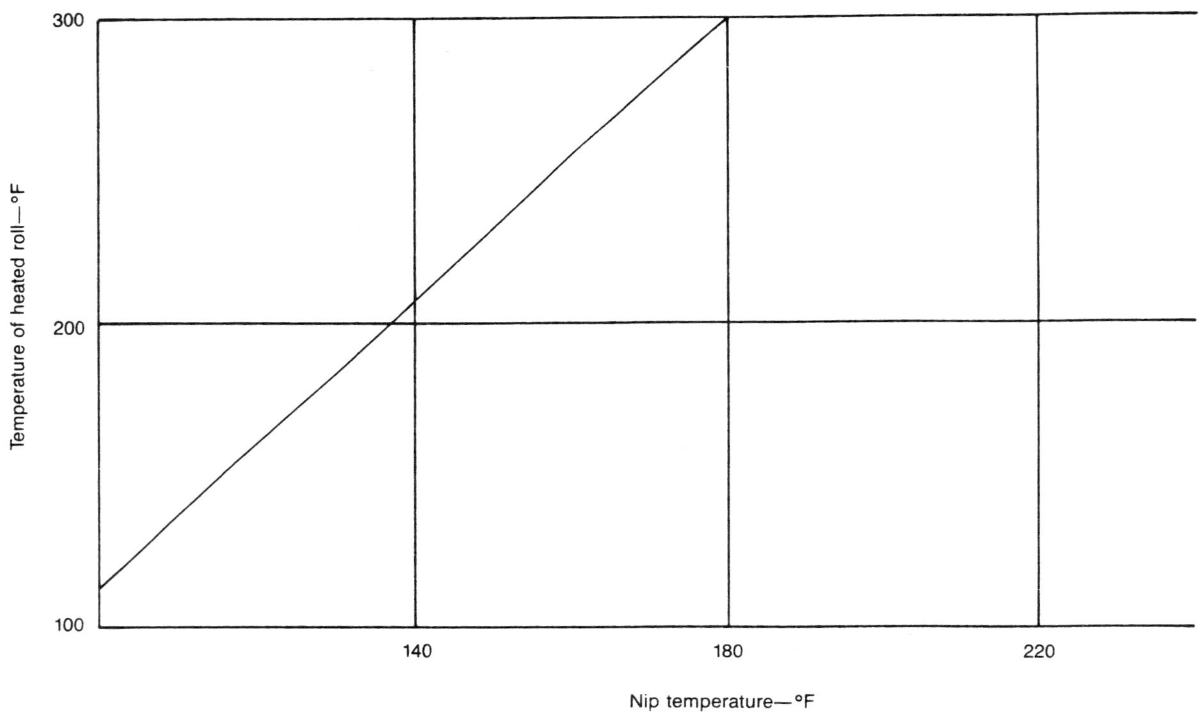

FIGURE 6.6. Heated roll temperature vs. nip temperature—no wrap.

spread between roll and nip temperature increases linearly as the roll temperature becomes higher – a not wholly unexpected result.

If the supplier cannot tell the converter what the activation temperature is, the converter can have his Q.A. laboratory determine it using a Packaging Industries heat seal, where pressure, temperature, and time of contact may be controlled. With this information, and knowing the configuration of the machine and the speed of the web, the converter will be able, at least, to compute a good place to begin.

CHAPTER 7

Unit Operation 6–Roll Winding and Slitting

EVERY converter salesman has had complaints about roll conformation from his customers. Every film salesman has heard similar tales from every converter he calls on, and the paper salesman is an old hand at coping with the assorted descriptions of deformed rolls he hears about. The complaints range from telescoping to crushed cores to wrinkles and buckles in the web that show up as the roll is unwound and the core is approached. And frequently, the wrinkles and buckles and creases are severe enough to cause that part of the roll of material to be unusable.

If the stock is reasonably uniform in thickness (plus or minus 10% of its specification) and if it contains no gauge bands, which create special winding problems of their own, then the defects in roll conformation are due to the use of an improper winding tension pattern.

HOW ROLLS ARE WOUND

There are two methods of winding rolls of film or paper or laminated stock: center and surface. The converter used center winding almost exclusively. The paper maker uses surface winding almost exclusively. In center winding, the core upon which the stock is to be wound is placed on a shaft with chucks holding it in position. The shaft is mounted on a mechanism that turns it, thus winding the stock attached to the core into a roll. In surface winding, the core, to which the end of the stock is attached, is placed in the crevice between two relatively large metal rolls parallel to each other and lying in the same horizontal plane, the size of the crevice depending on the diameter of the rolls and the distance between them. When the rolls are rotated, surface friction causes the stock to wind up on the core. This method is used for winding hard, large diameter rolls.

Any material wound on a core has tension applied to it in the act of winding it. Each layer of stock laid on the roll exerts a slight pressure directed toward the core resulting from the applied tension. If the tension in the wrap were zero, there would be no resultant pressure. But there is always tension in the wrap and there is always a resultant pressure.

WHAT EACH WRAP PRESSURE DOES

The pressure due to a single wrap is exerted on all the wraps below it and each layer of material laid on the roll adds to this pressure. The accumulated pressure forces all the lower wraps closer to the core, causing the circumferential space available to each wrap to be reduced. At the same time the compression causes each wrap to be reduced in thickness infinitesimally leading to an infinitesimal increase in length. Thus, while the circumferential space available for accommodating each wrap is decreased, the length of each wrap is increased. These effects result in a reduction in tension in each wrap. In the finished, wound roll, the residual tension in each wrap is not the tension applied to it when it was wound, but a different, somewhat lower one. The residual tension in any single wrap is dependent on the number of wraps above it, the winding tension applied to these wraps, the core diameter, and the diameter of the finished roll, in addition to some properties of the material constituting the wound roll.

If the pressure is great enough – and this occurs with moderate frequency, especially in larger diameter rolls – there will be a region as one goes into the roll from the outside where the wraps will become slack and where the space available for the wrap to occupy is smaller than the actual length of the wrap. The stock will tend to fold over on itself in order to fit into the space provided. The ends of such a roll present a starred pattern. Bands will appear, radiating from the core to the outer wraps of the roll, becoming less and less visible as they move farther from the core and, at some point, finally disappearing. Closer examination of these bands indicates the presence of buckles in a number of layers aligned one over the other, giving the overall impression of a band. These are arranged symmetrically around the roll resulting in a star pattern. Frequently, these buckles, folds, and wrinkles are so severe the material becomes waste.

WINDING PATTERNS

There have been many proposals made as to the proper way to wind a roll of any material. One is to maintain a constant tension. But this will lead to sufficient compression in the lower wraps, when a moderately large roll is wound, to cause a star pattern due to wrinkling and buckling of these lower wraps. One of the worst things done is to start winding at too low a tension and tighten the roll as it increases in size. This kind of error will most assuredly lead to buckles and wrinkles in the stock as the core is approached.

Another suggested pattern is to wind a roll at a constant torque. Put into mathematical terms, this means that $tr = A$, where t is the applied tension, r is roll radius at the wrap being wound and A is a constant. But such a pattern might lead to a very loose wind as the roll diameter increases. The outside wraps might slide off the roll. Also, such a pattern can lead to soft rolls, with a tendency to telescope.

A third proposal is to wind the roll in such a pattern as to yield a constant residual tension in the finished roll. This may be the most desirable of the three winding tension patterns proposed, but it is also the most difficult to do.

THE MATHEMATICS OF THE THREE PROPOSED PATTERNS

A detailed mathematical development of all these winding patterns was done for the winding of thread in 1962 by Margaret Catlow and G. W. Walls of the Cotton, Silk and Manmade Fibers Research Association located at Dudsbury, Manchester, England. While their analysis applies to winding of yarn, it is of sufficient generality to be adapted for use for anything wound on a core. Their results have meaning in the winding of substrates such as paper, plastic films, laminates, and even sheet materials, so long as they are wound in rolls on a core.

The results, adapted for winding of webs of material, for the three cases mentioned are shown below:

(1) Constant applied tension

$$T/t = 1 - [(r^2 - m^2)/2r^2] \ln [R^2 + m^2]/[r^2 + m^2] \qquad (7.1)$$

where:

T = residual tension in wrap
t = applied tension on wrap
r = radial position of wrap measured from center of core
R = outside radius of roll
m^2 = $a^2 (1 - p)/(1 + p)$ where p is Poisson's ratio for the material being wound, and a is radius of the core. (If a material is stretched longitudinally, then it shrinks laterally. Poisson's ratio is the ratio of the lateral shrinkage per unit width to the longitudinal stretch per unit length.)

(2) Constant torque

$$T/t = 1 - [(r^2 - m^2)/mr] \arctan [(Rr - r)/(m + Rr/m)] \quad (7.2)$$

(3) Constant residual tension

$$T/t = r(r^2 + m^2)/[r(r^2 + m^2) + (R - r)(r^2 - m^2)] \quad (7.3)$$

In all three cases, the ratio of residual tension to applied tension depends on a single characteristic of the material (Poisson's ratio) and the core and roll radii. The residual tension at each radial position within the roll is dependent on these three numbers and the applied tension, a relatively simple relation. In the derivation of these equations, the condition was made that in no case did the applied tension create a stress exceeding the yield point (elastic limit) of the material being wound.

In Tables 7.1 and 7.2, the value of the ratio of applied tension to residual tension for various values of the radial position of the wrap for constant applied tension.

These calculations have been made for a 1.75″ core radius (3.5″ O.D.), a 3″ radius core, a 7″ radius roll and 12″ radius roll; four different

TABLE 7.1. Ratio of Residual Tension to Applied Tension as a Function of Radial Position of Wrap.

Constant Applied Tension—3.5″ O.D. Core		
Radial Position of Wrap	T/t 7″ Radius Roll (14″ O.D.)	T/t 12″ Radius Roll (24″ O.D.)
1.75	0.406	0.143
2.0	0.321	−0.004
2.5	0.296	−0.089
3.0	0.351	−0.087
4.0	0.522	0.044
5.0	0.698	0.202
6.0	0.858	0.352
7.0	1.0	0.490
8.0		0.611
9.0		0.722
10.0		0.823
11.0		0.914
12.0		1.0

TABLE 7.2. Ratio of Residual Tension to Applied Tension as a Function of Radial Position of Wrap.

Constant Applied Tension—6″ O.D. Core		
Radial Position of Wrap	*T/t* 7″ Radius Roll (14″ O.D.)	*T/t* 12″ Radius Roll (24″ O.D.)
3.0	0.659	0.406
3.5	0.635	0.313
4.0	0.608	0.291
5.0	0.757	0.331
6.0	0.878	0.432
7.0	1.0	0.537
8.0		0.641
9.0		0.740
10.0		0.832
11.0		0.919
12.0		1.0

conditions for each of the winding patterns demonstrating the effect of roll and core size on *T/t*.

Reserving comment for the moment, a table of calculations involving a 6″ O.D. core for both sizes of roll is shown in Table 7.2.

The radial position is measured from the center of the core outwards. The radial position of the core surface, on which winding begins, is 1.75″ for one core size and 3″ for the other. The calculations were made assuming a value of 0.33 for Poisson's ratio, a value for many materials.

The tables show that the residual tension is less than the applied tension at every wrap position and equal to it only at the roll surface. Since the applied tension is held constant, the ratio of applied tension to residual tension represents the fraction of applied tension that is left in the wrap. Each successive wrap going outward from the core, with the exception of the first few, has a residual tension/applied tension ratio greater than the wrap immediately below it. Since the applied tension is held constant, this ratio represents the fraction of applied tension left as residual tension in each wrap. Thus, the residual tension increases as the radial position of the wrap increases. Since the applied tension is not reduced as the roll increases in radius, all the lower wraps will be slack. In the case of a 1.75″ core and a 24″ O.D. roll some of the lower wraps display negative tension, meaning, theoretically, that they are under a vertical compression.

Such rolls will unquestionably develop buckles and wrinkles and a star pattern on their ends. Holding the applied tension constant is the wrong way to wind a roll. Obviously, a method of winding which incrementally reduces the applied tension as the roll increases in size is a much wiser way to wind a roll. Increasing the tension at any point during the wind-up may lead to disastrous consequences in roll conformation.

Thus, the radial position of the core surface (on which winding begins) is 1.75″ for the 3.5″ O.D. core and 3″ for the 6″ O.D. core, two sizes in common use among converters and their suppliers. The calculations were made assuming Poisson's ratio to be equal to 0.33, a value found for a large number of materials.

Tables 7.3 and 7.4 show the same calculations made for a constant-torque, applied-tension winding pattern. This means, in mathematical terms, that $tr = A$, a constant. This is the equation of an equilateral hyperbola with the maximum applied tension at the smallest radius, which is at the core.

Since the applied tension follows a curve downward where $t = A/r$, the reduction and then rise in the applied tension/residual tension ratio does not have the significance that it does for a constant applied tension.

For a constant-torque winding pattern, there are no negative ratio values. Since, as the roll increases in diameter, the applied tension decreases, this may be close to a suitable winding tension pattern.

The last winding tension pattern to be considered is one which yields a constant residual tension. This is probably the ideal way to wind a roll, and the most difficult method to use. However, with today's programmable

TABLE 7.3. Ratio of Residual Tension to Applied Tension as a Function of Radial Position of Wrap.

Constant Torque—3.5″ O.D. Core		
Radial Position of Wrap	*T/t* 7″ Radius Roll (14″ O.D.)	*T/t* 12″ Radius Roll (24″ O.D.)
1.75	0.691	0.619
3.0	0.566	0.420
4.0	0.633	0.424
5.0	0.743	0.470
6.0	0.863	0.533
7.0	1.0	0.604
8.0		0.681
9.0		0.758
10.0		0.838
11.0		0.919
12.0		1.0

TABLE 7.4. Ratio of Residual Tension to Applied Tension as a Function of Radial Position of Wrap.

Constant Torque—6″ O.D. Core		
Radial Position of Wrap	*T/t* 7″ Radius Roll (14″ O.D.)	*T/t* 12″ Radius Roll (24″ O.D.)
3.0	0.775	0.691
4.0	0.739	0.578
5.0	0.767	0.564
6.0	0.888	0.593
7.0	1.0	0.635
8.0		0.705
9.0		0.774
10.0		0.847
11.0		0.922
12.0		1.0

slitter-winders, it becomes feasible. Calculations are shown in Tables 7.5 and 7.6.

The data in Tables 7.5 and 7.6 are only for the applied tension/residual tension ratio in the individual wrap at the radial position referred to. They take into account that the applied tension may have been reduced.

Demonstrating how to use the information provided in these tables may best be done by using an actual example of winding with a defined applied tension.

A Proposal for Determining a Suitable Maximum Applied Winding Tension

A judicious use of Hooke's law may be the best way to determine the magnitude of the maximum applied tension. Regardless of the winding pattern used, a method for specifying the maximum applied tension at the start of winding, or at the point in the winding operation where the maximum applied tension is reached, is a useful device.

Hooke's law states that $S = Ee$, where S is the stress producing strain e. But $S\delta = t$ where δ is web thickness and t is the applied tension. E is a constant called the modulus of elasticity. Then, $t/\delta = Ee$ or $t = Ee\delta$, or, in English, applied tension equals tensile modulus of elasticity multiplied by the strain produced by the applied stress, and this product, in turn, multiplied by the thickness of the substrate.

TABLE 7.5. Ratio of Residual Tension to Applied Tension as a Function of Radial Position of Wrap.

Constant Residual Tension—3.5″ O.D. Core		
Radial Position of Wrap	*T/t* 7″ Radius Roll (14″ O.D.)	*T/t* 12″ Radius Roll (24″ O.D.)
1.75	0.503	0.391
2.50	0.479	0.304
3.0	0.515	0.320
4.0	0.618	0.371
5.0	0.739	0.449
6.0	0.867	0.521
7.0	1.0	0.598
8.0		0.678
9.0		0.757
10.0		0.838
11.0		0.910
12.0		1.0

TABLE 7.6. Ratio of Residual Tension to Applied Tension as a Function of Radial Position of Wrap.

Constant Residual Tension—6″ O.D. Core		
Radial Position of Wrap	*T/t* 7″ Radius Roll (14″ O.D.)	*T/t* 12″ Radius Roll (24″ O.D.)
3.0	0.694	0.503
3.5	0.685	0.473
4.0	0.715	0.472
5.0	0.783	0.508
6.0	0.886	0.563
7.0	1.0	0.628
8.0		0.697
9.0		0.770
10.0		0.845
11.0		0.923
12.0		1.0

Table 7.7 shows the value of the tensile modulus of elasticity for a number of plastic films.

To compute an approximate modulus for a composite, a weighted average may be taken. For a 0.0005 PET combined with 0.002 PE, we have 80% PE and 20% PET, based on thickness of the films. This gives $0.8 \times 25{,}000 + 0.2 \times 500{,}000$ for a total of 120,000. Values of tensile modulus vary widely among materials used by the converter. It is used as a measure of stiffness; the higher the modulus, the stiffer the material.

Next, the maximum applied tension must be determined. A method for arriving at this must be developed that can be generalized to apply to whatever substrate the converter runs.

A Proposal for Determining a Suitable Maximum Applied Winding Tension

A judicious use of Hooke's law may be the logical way to proceed to select a useful maximum winding tension. If it is assumed that a suitable maximum applied tension may be arrived at by taking a tension that produces a given strain, then knowing the thickness of the web and its modulus of elasticity, a maximum applied tension may be computed.

Let it be assumed that the tension producing a 1% strain in the PET/PE laminate mentioned above is chosen as the maximum applied tension to be used. Since it has already been demonstrated that a constant applied tension pattern is a bad way to wind a roll, this pattern will be ignored and calculations will be made only for the constant torque and constant residual tension patterns to construct curves showing both the applied tension and resulting residual tensions.

For a constant torque pattern, the maximum applied tension is at the core. It falls off from this r following the equation $tr = A$. The maximum applied tension may be found from the equation $t = Ee\delta = 120{,}000 \times 0.01 \times$

TABLE 7.7. Tensile Modulus of Elasticity for Various Plastic Films.

Film	Tensile Modulus
PE (density—0.918)	25,000
PE (density—0.950)	110,000
Polypropylene (oriented)	170,000
Nylon 66	330,000
Polyethylene terephthalate (oriented)	500,000

Source: Harper, C. A., ed. 1975. *Handbook of Plastics.* New York: McGraw-Hill. See also manufacturer's (Instron Corporation) literature.

0.0025 = 3 lbs./inch of width. Since this is applied at the core, 3 × 1.75 = 5.25, or $A = 5.25$.

Thus, the equation for the applied tension curve is $tr = 5.25$.

It now is possible to find the points on the applied tension curve and, with the aid of the residual tension/applied tension ratios, calculate the resulting residual tension.

These results are shown in Table 7.8.

At no point is the residual tension in any wrap less than that in the wrap above. The danger of buckles and wrinkles in the lower wraps is non-existent, but the residual tension is reduced to 0.5 lbs/inch of width at a radial position of 5″ and drops marginally further to 0.4375 at the roll surface.

The wraps in the top 7″ of the roll may be held too loosely and may tend to telescope, if a high slip film is wound. Or, the roll may be too soft for easy handling.

This may be overcome by increasing the maximum applied tension to a higher level and still remain well below the elastic limit of the material. It might be considered that the minimum residual tension should be 1 lb. per inch of width. The point of minimum residual and applied tension is at the roll surface or at a radial position of 12″, where the two are equal.

Therefore, $tr = 1 \times 12 = A$, and the new equation of the applied tension curve is $tr = 12$. At the core, where the maximum applied tension occurs, $t = 6.857$, equivalent to a 2.286% strain, not enough to produce a crushed core or damage the film.

A constant torque applied winding tension pattern may produce satisfactory rolls with the proviso that the minimum applied tension be high enough to prevent telescoping when a high slip film is wound.

TABLE 7.8. Applied Tension and Residual Tension as a Function of Radial Position of Wrap.

Constant Torque—1% Strain		
Radial Position of Wrap	Residual Tension	Applied Tension
1.75	1.857	3.00
3.0	0.736	1.751
4.0	0.556	1.313
5.0	0.493	1.05
6.0	0.466	0.875
7.0	0.453	0.75
8.0	0.442	0.656
9.0	0.442	0.583
10.0	0.449	0.525
11.0	0.4385	0.477
12.0	0.4375	0.4375

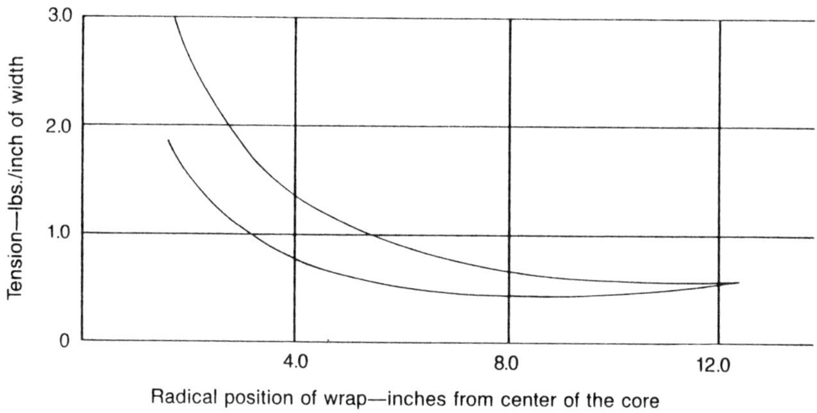

FIGURE 7.1. Tension in wrap vs. wrap position—constant torque.

The calculations shown in Table 7.8 are plotted in Figure 7.1.

While an applied winding tension pattern that results in a constant residual tension may be the ideal way to wind a roll, even with today's programmable machines, it may still be difficult to do.

The procedure used to perform the calculations was first to determine the location of the maximum applied tension. The value is known to be, for a 1% strain, 3 lbs./inch of width. Now to find the radial position. This can be done in two ways: the elegant one of differentiating Equation (7.3) with respect to r, setting the resulting $d[T/t]/dr = 0$ and solving for r, or the brute

TABLE 7.9. Applied Tension and Residual Tension as a Function of Radial Position of Wrap.

Constant Residual Tension—1% Strain		
Radial Position of Wrap	Residual Tension	Applied Tension
1.75	0.9022	2.646
2.32		3.00
3.0		2.819
4.0		2.432
5.0		2.009
6.0		1.732
7.0		1.509
8.0		1.330
9.0		1.182
10.0		1.077
11.0		0.082
12.0		0.9022

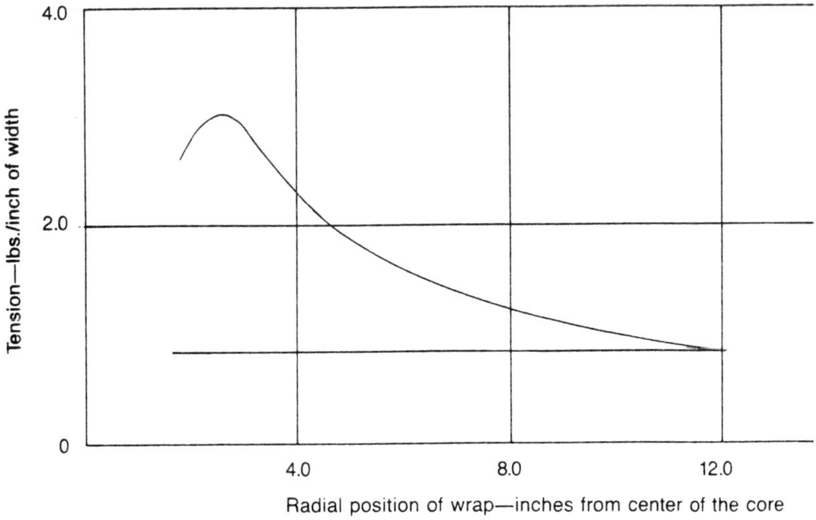

FIGURE 7.2. Tension in wrap vs. position of wrap—constant residual tension.

force method of incremental calculations. In any case, both methods gave the result of 2.32″ and residual tension/applied tension ratio of 0.300719971. The residual tension at the point of maximum applied tension is 3 × 0.300719917 = 0.9022.

Since the residual tension is constant, applied tension at any point may be found by dividing 0.9022 by the residual tension/applied tension ratio at each wrap for which it is desired to find a value. Table 7.9 shows these calculations.

Since 0.9022 lbs./inch of width may be sufficient tension to wind a hard enough roll, avoiding both a loose wrap or softness, a 1% strain may provide adequate tension for producing satisfactory rolls, regardless of how high the slip level of the material being wound is.

Table 7.9 is plotted in Figure 7.2.

Any applied winding tension pattern between a constant torque and a constant residual tension winding pattern would give a satisfactorily wound roll, with the proviso for a constant torque pattern already mentioned.

HOW STOCKS ARE CUT

There are three methods used to cut the contents of a roll to specified sizes: score cutting of the web, scissor cutting of the web, and slicing through the wound roll with a rotating circular blade. This latter is not used by converters because of their extensive use of thermoplastic films and coatings. With these materials, the edges of layers are in danger of being welded together due to the frictional heat generated. It is used in the paper industry for the cutting rolls of tissue and paper towels.

Score cutting is accomplished by having a circular knife ride against a hardened steel roll. The stock to be cut is fed between the knife and the roll and the cut stock is rewound. If stock is being trimmed, the trim is taken away by a suction pipe. If rolls are being wound up for shipment, the width is determined by the distance between knives. The knives, of course, are held on a shaft designed for that purpose.

Scissor cutting is done by having the knives run in a groove on a hardened steel roll, providing a scissor action between the knife and the edge of the groove. Since space must be provided on the roll for the grooves, scissor cutting does not give the infinite variability in width available in score cutting. This defect has not eliminated its use. It is well adapted for cutting soft, extensible films. Score cutting can be used for this same purpose, with the proviso that the knives be kept sharp.

There is some discussion among the cognoscenti concerning the angle at which a knife blade should be sharpened. Since there are many materials to be slit and many widths, including ribbons and yarn, where scissor cutting is used exclusively and the knives are flat with highly finished edges equivalent to a 90° angle, the discussion has led to different angles being used for different purposes. The converter may select a procedure for winding somewhere between a constant torque applied tension winding pattern and a constant residual tension pattern, and using the application of Hooke's law suggested in the text, determine the maximum applied tension he wishes to use.

Properly wound and slit rolls are important to the customer. Being able to run right down to the core is pleasing. With selection of the proper winding tension pattern and a proper maximum applied tension, the converter will be able to supply his customers with properly wound rolls consistently.

CHAPTER 8

Unit Operation 7 – Heat Sealing

MANY of the materials used by the converter are thermoplastic polymers which are weldable with the use of heat and a modicum of pressure. Making such welds is known as heat sealing and is a common method of forming pouches and closing overwraps. Not every thermoplastic polymer is heat sealable: cellulose acetate and thermoplastic polyesters are good examples of this type of polymer. There are many polymers which can be formulated into coatings that serve as heat-sealable closure materials and, of course, there is PE, which can be heat sealed into a bread bag by itself, or find a whole host of uses as the weldable inside layer of a multitude of laminations.

Thus, the final unit operation related to converting for flexible packaging is the act of heat sealing, which is performed when the converter is providing a customer with a pouch made of whatever material the customer has ordered; PE for a bread bag, PET laminated to polypropylene for a hospital package, or other material with a heat-sealable surface for whatever purpose. Since form-fill-seal machines for many products, from potato chips to paper goods, are presently available, pouches are being ordered less frequently and orders for roll stock are common.

Nevertheless, most converters have pouch machines of one kind or another and there are many varieties ranging in size from machines that will make very small pouches to machines that will make large ones. The principles of operation are the same – the heat seals are made either by clamping a pair of jaws over the heat-sealing material, one or both of which are heated, or by running the heat-sealing web between a pair of wheels one or both of which are heated. The process might be called conduction heat sealing, because heat is conducted from a heated bar through the web to the heat-sealing surfaces.

A pouch machine is essentially a form-fill-seal machine without the filling mechanism. Many are single-web machines, although some, which manufacture large bags, use two webs of material. If front and back must be registered, it is much easier to do this by folding over a printed web than by feeding two webs. Registration of two webs can be done, but the means are relatively costly and complex.

Nevertheless, almost all converters have pouch equipment, perhaps dating back to the days when they supplied pouches to almost all their customers.

HOW A POUCH MACHINE OPERATES

Consider a single web machine. A roll of stock is mounted on one end and unwound over a former. The web is folded so that the heat-sealing layer faces itself as the inside of the folded web. The printed legend reads perpendicularly to the direction of travel of the web through the machine. A pair of driven rolls feed the web into the heat-sealing jaws, which are perpendicular to the direction of web travel. These are brought together by a cam. Following the heat seal, the web is moved forward by a pair of driven rolls to where the seal is cooled and cut through by a knife. Registration is often maintained by underfeeding the web slightly, and stretching it so that the cutoff occurs in the same place. Electric eye spots are printed on the web and control of cutoff is sustained by the action of the photocell. The mechanism of a commercial pouch machine is well known, and there are many manufacturers of pouch-making machines, such as Doboy Packaging Machinery, FMC Packaging Systems Development, Wrap-Ade Machine Co., and others.

HEATED JAWS FOR SEALING

Consider first the case where both jaws of the heat-sealing mechanism are heated to the same temperature. This is shown schematically in Figure 8.1. It is a configuration sometimes used. The thin slab of heat-resistant rubber is placed over one of the jaws to accommodate the defects in the surface of the other. When the sealing layer is thick and a high temperature at the jaw surface is needed to make a satisfactory seal, one or both of the jaws of the heat-sealing mechanism are often covered with a sheet of Teflon-coated or silicone-coated fiber glass to prevent sticking of the web.

The equation describing the distribution of temperature in the web under these conditions is one that has been seen in this work before:

$$[T_j - T]/[T_j - T_a] = [4 \sin p/(2p - 2 \sin 2)]\, e^{-p^2(at/2w)^2} \cos px/w \qquad (8.1)$$

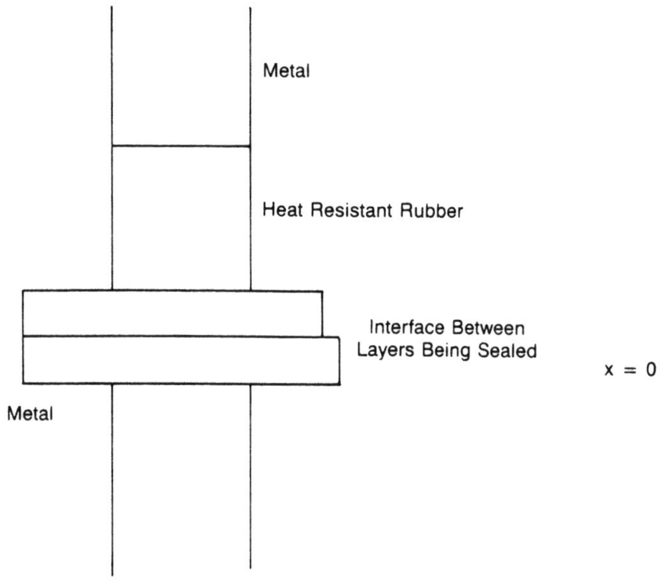

FIGURE 8.1. Both jaws heated.

where:

T_j = platen or jaw temperature
T = temperature at interface of the two layers being sealed
T_a = ambient temperature
p = a root of $p \tan p = C$ [see Chapter 6, Equation (6.7)]
$2w$ = thickness of web being sealed
a = thermal diffusivity of web, cm squared/sec
t = time coordinate
x = coordinate parallel to direction of flow of heat, with origin at heat-sealing interface

If Equation (8.1) is converted into a logarithmic form, it is a little easier to manipulate.

$$\ln [T_j - T]/[T_j - T_a] = A - p^2 (at/2w^2) \ln \cos px/w$$
$$A = \ln [4 \sin p/(2p + 2 \sin 2p)] \tag{8.2}$$

Since there is no convection, taking the convective heat transfer coefficient as infinite gives $p = \pi/2$. Then, $\sin p = 1$, $\sin 2p = 0$, making $A = 0.243$. Taking x as equal to 0, which is the location of the interfaces being sealed, Equation (8.2) becomes:

$$\ln [T_j - T]/[T_j - T_a] = 0.243 + \ln \cos 0 - p^2 at/w^2 \tag{8.3}$$

Simplifying further gives:

$$\ln[T_j - T]/[T_j - T_a] = 0.243 - 83.125\, t \tag{8.4}$$

(Note: p has been defined in the chapter on coating.)

It is possible to calculate the time required for the interfacial temperature to reach a given value if the jaw temperature and ambient temperature are given.

As an example, if it is assumed that the jaw temperature is 400°F and the ambient air temperature is 80°F; and it is also assumed that a laminate of PET to PE with an acrylic adhesive is being sealed, the physical properties of the materials involved needed to make the calculations are:

PET
Density = 1.40 gm/cc
Thermal conductivity = 5.8×10^{-4} cal/sec·cm·°C
Specific heat = 0.4 cal/gm·°C
Thickness = 0.0005 inches = 1.27×10^{-3} cm
Acrylic adhesive: assume 1 lb/ream of dry adhesive
Density = 1.17 gm/cc
Thermal conductivity = 1.4×10^{-4} cal/sec·cm·°C
Specific heat = 0.35 cal/gm·°C
Thickness = 1.391×10^{-4} cm

PE
Density = 0.915 gm/cc
Thermal conductivity = 8.0×10^{-4} cal/sec·cm·°C
Specific heat = 0.55 cal/gm·°C
Thickness = 0.002 inches = 5.08×10^{-3} cm

Instead of calculating time for heat to pass through each layer of the structure, which is the precise way; an approximate set of properties for the composite structure may be computed by taking a weighted average for each property. For example, to find the density of the composite, we take: for PET, $1.40 \times 1.27 \times 10^{-3} = 1.778 \times 10^{-3}$; for PE, $0.915 \times 5.08 \times 10^{-3} = 4.648 \times 10^{-3}$; and for the acrylic adhesive, $1.17 \times 1.391 \times 10^{-4} =$

1.627×10^{-4}. Adding it all up gives 6.581×10^{-3}. Divide by the total thickness and we get 1.015 as the composite density. These calculated composite properties are shown below:

Density = 1.015 gm/cc
Thermal conductivity = 7.428×10^{-4} cal/sec·cm·°C
Specific heat = 0.516 cal/gm·°C
Thickness = 6.489×10^{-3} cm

And from the above, a value for the thermal diffusivity may be calculated, yielding:

Composite thermal diffusivity = 1.418×10^{-3} cm²/sec

One may now proceed with computing the time needed, given the conditions stated, to reach a given temperature at the interface between the two surfaces being sealed together, bearing in mind that a minimum heat sealing temperature for PE is 220°F. A set of time and temperature values is shown in Table 8.1.

The time required for the heat-sealing surfaces to reach the designated temperature for a structure consisting of a 0.0005″ PET laminated with 1 lb/ream of an acrylic adhesive to 0.002″ PE is of the order of 10^{-2} sec. If the machine is operating at 100 pouches/min., this is 0.6 seconds for a complete cycle of the machine, far more time than is needed for the surfaces to reach the postulated temperature.

TABLE 8.1. Time Required to Reach a Given Temperature at Heat-Sealing Interface—Both Jaws Heated to 400°F.

Temperature at Interface, °F	Dimensionless Temperature Difference Ratio	Elapsed Time to Reach Temperature, sec × 10^2
230	0.53125	1.0537
240	0.5	1.1286
250	0.46875	1.2043
260	0.4375	1.2873
270	0.40625	1.3765
280	0.375	1.4728
290	0.34375	1.5776
300	0.3125	1.6923
320	0.250	1.960
340	0.1875	2.306

However, more is required than merely heating the surfaces to make a heat seal. The behavior of the molecules of the heat-sealing layer of polymer under the conditions of temperature, pressure, and duration of contact of the two heat-sealing layers has more to do with seal quality than time required to reach temperature.

Figure 8.2 shows another configuration, with the lower jaw (or platen), unheated. The lower jaw consists of a chunk of heat-resistant rubber machined to the proper dimensions and fitted into the machine. Silicone rubber, and sometimes compounded neoprene or other formulated elastomers, are used to overcome the mechanical defects of the unyielding metal surface. This is a far more popular configuration than that shown in Figure 8.1. Sometimes the rubber jaw is heated to a lower temperature than the metal one.

Under these circumstances, the equation showing temperature distribution becomes:

$$[T_j - T]/[T_j - T_a] = \text{erf}\, x/2(at)^{1/2} \tag{8.5}$$

where:

T_j = temperature of upper jaw
T = temperature of interface between two layers being sealed
T_a = ambient air temperature
w = thickness of web
a = thermal diffusivity of substance of which web is made
t = time coordinate
x = coordinate in direction of heat flow
erf = designation meaning "error function," values of which are given in any good set of mathematical tables, one of which is the *Handbook of Mathematical Functions,* publication #55 of Applied Mathematics Section of the National Bureau of Standards

Zero for the x-axis is at the outside surface of the web at the unheated jaw.

Using data already given, and the values of the error function found in a set of tables, it is possible to compute a set of time and temperature values similar to those in Table 8.1, as shown in Table 8.2.

What is most striking about the calculated values of time is their remarkable brevity. With both jaws heated, their general order of magnitude is 10^{-2} seconds. With one jaw heated, the general order of magnitude becomes $10^{-1.5}$ seconds, indicating that the time to reach a heat-sealing temperature is not necessarily the controlling factor in the time required to make a seal.

As already indicated, most pouch machines do not run faster than 100

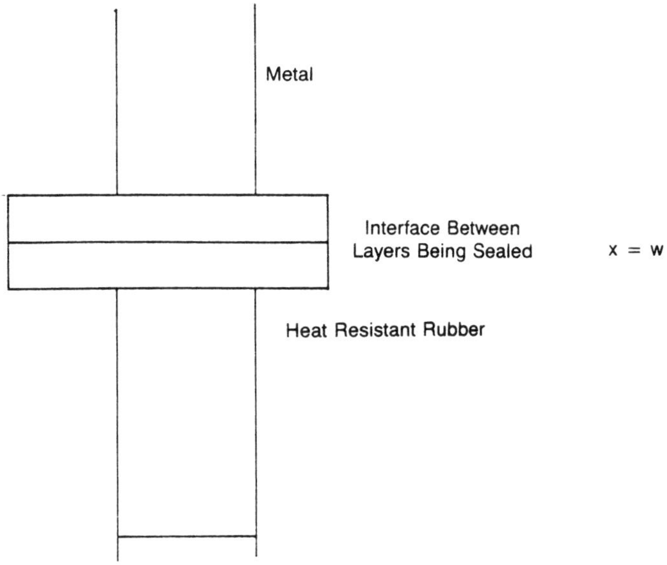

FIGURE 8.2. One jaw heated, the other neutral.

cycles per minute. Those making large bags and a multiple number with each cycle run more slowly.

But consider 100 cycles per minute as a good number to select as a point of discussion. At 100 cycles per minute, 0.6 seconds is permitted for making the heat seal.

TABLE 8.2. Time Required to Reach a Given Temperature at Heat-Sealing Interface under the Conditions Shown in Figure 8.2.

Temperature, °F	Dimensionless Temperature Difference Ratio	Elapsed Time to Reach Temperature, sec × 10^2
230	0.53125	2.476
240	0.5	2.740
250	0.46875	3.059
260	0.4375	3.450
270	0.40625	3.930
280	0.375	4.465
290	0.34375	5.337
300	0.3125	6.538

Graphs of the data in Tables 8.1 and 8.2 are shown in Figures 8.3 (both jaws heated) and 8.4 (one jaw heated).

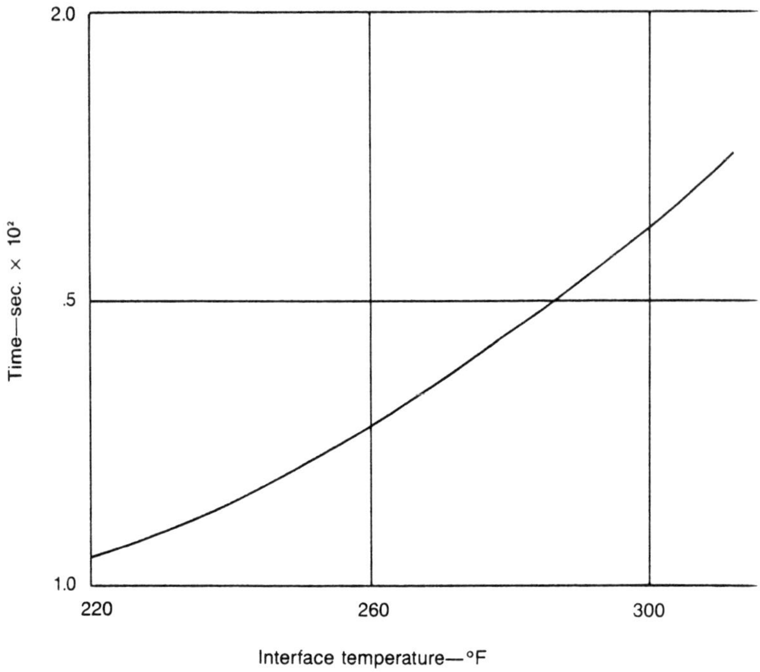

FIGURE 8.3. Time for interface to reach a given temperature. Both jaws heated to 400°F vs. given temperature.

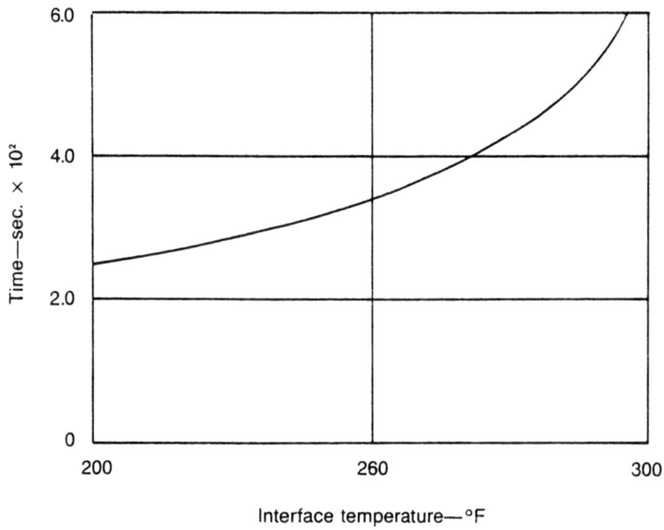

FIGURE 8.4. Time for interface to reach a given temperature. One jaw heated to 400°F. Second jaw at ambient conditions vs. given temperature.

This is much more time than the time needed to reach a heat-sealing temperature at the heat-sealing interface. The time needed to execute a heat seal is mostly the time needed to cause the heat-sealing material to flow and, with the combination of applied heat and pressure, to weld into a single layer. It is possible that some small portion of the heat-sealing material is extruded out of the sealing area into the interior of the pouch causing thinning of the web in the seal. It is also possible that the material comprising the sealing area is slightly compressed, adding to the thinning of the web. It has been found that the inside edge of the heat seal is often the weakest part of the pouch, perhaps due to the apparent thinning of the seal.

It is also quite likely that a further cause for the length of time required to make a heat seal is that the web carries away a good deal of surface heat, and that time is required to bring the surface back to temperature. Instead of putting the temperature control along the outside of the jaw, the sensor is placed into the jaw in such a way as to be at the surface, where it is ground smooth and made a part of the heat-sealing jaw itself. While this gives a rapid response of the jaw heater, it does not accelerate the time for heat to travel from its interior to its surface.

If the area of the heat seal is taken as 25.4 cm^2, with the web in the example chosen having a thickness of 6.489×10^{-3} cm, a volume of 0.33 cc is obtained. With a composite density of 1.015, this yields a mass of 0.335 gm. Multiplying by a specific heat of 0.516 and a temperature change of 177.8°C (equivalent to 320°F) gives 51.4 cal. as the amount of heat absorbed by the web in being heated from 80°F to 400°F.

Now the problem of how thick a surface is arises. Is it a micron or a millimeter? Is it a monomolecular layer or is it a number of layers? If it is assumed that the sealing bar is one-inch thick and made from aluminum, and that the effect of the web absorbing 51.4 calories from it is to cool the entire mass, then the cooling effect is 51.4/230 or 0.22°C, a negligible amount. An approach to this problem could be made by calculating the heat flux from the surface of the internal bar heat source, assuming a cylindrical shape of a given length and assuming a given surface temperature of the heating element.

Without some definition of how deep a surface may be considered to be, the problem is indeed difficult. Suffice it to say that the length of time required to make a true, welded seal is somewhat longer than the calculated time required to heat the web to a sealing temperature.

Suffice it also to say that the length of time required to make a truly welded seal is considerably longer than the calculated time required to heat the web to a sealing temperature if no cooling of the sealing jaws occurs.

Assuming, as above, a bar 10″ long, 1″ thick, and 1″ wide, we have a volume 10 in^3 and a surface area of 10 in^2. This is equivalent to 163.9 cm^3 and 64.2 cm^2. Considering a number of thicknesses of surface, we can calculate the reduction in surface temperature produced by heat extraction of the web, as shown in Table 8.3.

The thinner the surface, the greater the heat loss produced by the 51.4 calories required to raise the web temperature from 80°F to 400°F. This is not unexpected, since the mass of a thinner surface is lower than that of a thicker one. By the same token, it will reheat more rapidly. Nevertheless, it is not a factor to be ignored.

There are other methods of heat sealing. Dielectric sealing is one and hot wire sealing (sometimes called impulse sealing) is another. Dielectric sealing can be performed only on those materials having high dielectric dissipation factors. PE, by far the most common heat-sealing material, is not one of these. As for impulse sealing, it results in a thickened line seal and is satisfactory for many, many products. Unfortunately, it is difficult to adapt to a high-speed machine. Pouch machines and form-fill-seal machines generally use conduction heat sealing.

Dielectric sealing is used for making packages from PVdC film, since it works far better for this particular film than conduction heat sealing. Heating dielectrically does not involve hot jaws. The film may be heat-sealed while the jaws remain relatively cool. But only few polymers have the electrical characteristics required.

Heat sealing is one of the oldest arts in packaging. Over the years, there has been a steady development of efficiency, but little imaginative thought has been devoted to the process.

EPILOG

This completes the discussion of the unit operations of converting. Much more might have been said about each topic, but the essentials have been related in a way that, it is hoped, has not been dull.

TABLE 8.3. Temperature Loss in Surface of Aluminum Bar as a Function of Thickness.

Surface Thickness, cm	Temperature Reduction, °C
0.10	12.6
0.07	17.9
0.05	25.1
0.04	31.4
0.025	50.2
0.01	125.0

These operations represent the core of the converting industry. Engineering details have been left out, since much of the machine design is not germane to this work. Rather, the physical problems presented by the operation have been the subject matter discussed here.

The balance of the book is concerned with topics of interest to converters but which are ancillary to the converter's activities.

SECTION TWO

ANCILLARY CONVERTER CONCERNS

CHAPTER 9

Formulating

THE primary art related to converting as it has been described and expanded upon in these pages is one most converters don't do. In printing, coating, laminating and often in heat sealing, an ink, adhesive, or a coating to prevent scuffing or impart slip to a surface or provide a heat seal is needed for the finished package. There is an entire industry supplying these mixtures of materials to converters.

In the U.S. most converters purchase their inks, coatings and adhesives. Only a few make their own. In Europe, most converters make their own inks, coatings and adhesives, while a few purchase them. This structure has strong historical roots.

There are in the U.S. some converters that formulate their own inks, coatings and adhesives and farm them out for manufacture.

Making or formulating these compositions is based on little science and much art. A good formulator is a relative rarity, and once an organization has one, it is very reluctant to give him up. Most formulators are trained chemists, but a chemical background is not a necessity.

The composition of many formulations, as this class of materials is commonly referred to, is relatively simple, with some exceptions. The reaction of knowledgeable people, on learning of some competitor's formulation is likely to be "Of course, nothing to it," but it probably took many trials to get the desired result.

To formulate well, one must have a thorough knowledge of the characteristics of polymers, resins, resin additives, pigments, wetting agents, solvents and a goodly number of other materials which might be used in relatively small quantity, such as anti-blocking agents, slip agents, mold and bacterial inhibitors, et al. All of the materials used must conform to the FDA Title 1, Chapter 21 regulation 175.105, covering materials which may be used in "indirect contact with foods" such as an adhesive for laminating or an ink which will be reverse printed.

If used in direct contact, it should conform to the extraction regulations or to some specific regulation relating to the particular polymers used in the composition.

If the machinery on which the material will be run is equipped with incinerators, the formulation may use organic solvents. If not, it will require the use of water dispersions or 100% solids materials in order to conform to Environmental Protection Agency requirements or perhaps state environmental requirements – e.g., California.

Various government agencies have placed constraints on what the formulator may use and how much of it. While he still has a great deal of freedom to select whatever substances he chooses, all the various regulations establish limits.

Below, illustrating how one goes about making such compositions, two examples, one in solvent and one in water, are shown. Both are, or have been, used commercially.

FORMULATING IN ORGANIC SOLVENT

Usually, a start is made by defining the problem. It begins with a customer's request, made to a salesman, who brings it back and deposits it with the development people. If the request is for a heat-seal coating for aluminum foil, the formulator will begin with one set of substances. If it is for PET, another will be required. If it is for PE or PP, still a third set will be needed. Or, perhaps the formulator has sufficient confidence in his skill to formulate around an ethylene-vinyl acetate or ethylene-ethyl acrylate copolymer for all of them if plasma treatment is available in the customer's plant.

Taking as an example a heat-seal coating for aluminum foil with incinerators available on the machine on which the coating is to be run, it is possible to use an organic solvent in the formulation. Since taste and odor are always a problem, high-boiling solvents are to be avoided. With aluminum foil as the substrate to be coated, a polymer with some free carboxylic acid groups might be a starting point as a film-forming base on which to build. An ethylene-acrylic acid copolymer, an ethyl acrylate-acrylic acid copolymer, or perhaps a vinyl chloride-vinyl acetate-maleic anhydride terpolymer with 1% or less of the maleic anhydride moiety might be suitable. Or, a carboxylated rubber such as neoprene or nitrile rubber might be a possibility, assuming blocking can be overcome.

The choice is likely to be the vinyl terpolymer. It has been in existence for a long time, is in reasonably good supply, and has been used since its invention for foil coatings. To make it heat-sealable at relatively low temperatures, it is necessary to plasticize the polymer by adding something to a solution of it that is compatible with it and is soluble in the solvents used to dissolve the polymer. High-boiling esters of phthalic acid are suitable for this purpose, and are used frequently in those instances where they meet FDA requirements. There are also high-boiling phosphate esters that are suitable for softening the polymer.

Another set of substances that are compatible with the vinyl terpolymer are the various grades of nitrile rubber available which form a true solid solution with vinyl chloride polymers and interpolymers, and add strength and flexibility to the coating.

Since it is likely that the plasticized terpolymer will be slightly tacky and might have a tendency to block (as the tendency of the coating to stick to itself or the back surface of the roll of supported aluminum foil to which it is applied, is called), a small amount of anti-blocking agent needs to be added to overcome this. Further, an agent that produces an amount of slip in the coating may be necessary to prevent the coated web from binding to formers on a pouch or form-and-fill machine.

Anti-blocking agents are frequently hydrated silicon or aluminum oxides of various particle sizes, and should be added from ground dispersions rather than stirred in. Care must be exercised in using the aluminum oxides, since some of them tend to be abrasive and will wear coating rolls and doctor blades. Slip agents are waxes of various kinds, including synthetic materials such as long chain amides (e.g., erucamide).

All the above must be blended into a low-boiling solvent. In this case, methyl ethyl ketone is a good choice, with toluene as a diluent. Since the toluene is the last substance to come off when the coating is dried, if hot toluene were not a solvent for the solids of the formulation (with the exception of the anti-block and slip agents), the coating would precipitate out, yielding a hazy, weak, non-adherent mass. Care must be exercised in making solvent blends to be certain that the last portion of the blend to evaporate when the coating is dried is a solvent for the solid polymers in the formulation.

Possibilities for the final composition are shown in Table 9.1.

The Santocel and Acrawax are milled into the formulation. That is, a concentrate is made of perhaps 8% Santocel and 5–6% Acrawax dispersed in a formulation of 20% vinyl chloride terpolymer and 10% nitrile rubber (35% acrylonitrile) and ball milled to disperse the materials better (or a dispersion may be purchased from a firm specializing in dispersion manufacture). The additives are added to the finished product in this form in proper quantity.

This is a device frequently used in formulating. Certainly, in the manufacture of inks, pigments are added from color concentrates. There is an entire

TABLE 9.1. Final Composition.

Ingredient	Percent
Vinyl chloride-vinyl acetate-maleic anhydride terpolymer (Union Carbide – Vinylite VMCH)	20.0
35% acrylonitrile-65% butadiene copolymer (nitrile rubber) (BFGoodrich – Hycar OR 35. Also Exxon Perbunan)	10.0
Silicon dioxide anti-block (Monsanto – Santocell)	1.0
Slip agent – there are many – Glyco Products Acrawaxes, for one	0.7
Solvents – methyl cthyl ketone	54.6
toluene	13.7

industry that has grown by manufacturing only dispersions, and quite a number of companies will make such compositions to the user's specifications.

The above represents how one might go about making a formulation in solvent, where incineration (or solvent recovery) is available at a customer's plant.

FORMULATING IN WATER

But, more frequently, neither incineration nor solvent recovery is available, and, in order to avoid EPA problems (and also to be a good corporate citizen and avoid polluting the atmosphere with solvent vapors) the customer demands a non-polluting composition that meets all the EPA regulations. Suppose he needs an adhesive to replace his current isocyanate/polyol combination.

Machines capable of running 100% solids materials are not too common in the converting industry, although there are a few.

The problem is one of formulating in water as a carrier to insure that the customer conforms to all possible environmental regulations including those of a state and local nature.

In addition to its film-forming capability, the adhesive should cure to be comparable with the isocyanate/polyol composition the customer has been using.

To begin, an acrylic water dispersion (or latex) might be selected to provide a modicum of adhesion and film-formation. Secondly, the acrylic dispersion should be non-ionic in character, giving the formulator a wider latitude in the selection of additives. There are a number of acrylic dispersions ranging from very soft to very hard that are non-ionic in nature.

To make a formulation that will cure somewhat, one could select a long oil alkyd resin, an isocyanate/polyol, or an isocyanate/polyamine combination, an epoxy/polyamide combination, a dibasic acid/polyamine combination, or even a dibasic acid/trihydric alcohol pair.

But many of these systems require temperatures above room temperature to effect a reaction and a curing oven is not always available at a customer's plant. Also, some split-off water in curing must be absorbed by the adhesive layer.

It is advisable to select a curing reaction that will go smoothly at room temperature and that does not split-off water. This leaves the epoxy/polyamine and the isocyanate combinations.

Making an isocyanate stable in water requires prepolymerizing with dimethyl propionic acid and dispersing in water with triethyl amine, something not included in FDA regulation 175.105. Hence, of all the possible curing reactions available, only the epoxy/polyamide seems to meet all the governmental requirements.

Which epoxy? Which polyamide?

If a start were made with the workhorse of the plastics industry, *Epon 828* (Shell Chemical Co.) as the epoxy resin, then selecting a reactive polyamide to go with it becomes a relatively simple matter. Since the reacted material must not be brittle, and at the same time have some heat resistance, the selection of *Versamid 125* (Henkel Corp.) to react with the Epon 828 seems a reasonable choice.

Since one requirement for the adhesive is that it be a one-part system, a way of preventing the reaction from occurring when the adhesive is packed and stored must be found.

It is well-known that reactive polyamides like Versamid 125, containing free amino groups (Versamid 125 has an amine equivalent weight of 163) can be easily dispersed in water solutions of acids. If a volatile acid acceptable to the FDA can be found, the reaction can be suppressed while the adhesive is in storage. On drying, when the volatile acid is driven off, the reaction is free to take place. The acid converts the amino groups to ammonium ion groups, which do not react with epoxy groups. But, on drying, the ammonium ion group is converted back to an amino group and the reaction proceeds. It is not difficult to calculate how much acid to add, since the hydrogen equivalent weight of Versamid 125 is 100.

The ratio of Versamid to Epon 828 (which has an epoxy equivalent weight of 189) is 163/189.

Acetic acid is allowed by the FDA, and is volatile enough to suffice as the acid suppressant. The ratio of Versamid 125 to acetic acid (molecular weight = 60) is 100/60.

All the facts needed to formulate an acrylic adhesive with an epoxy/polyamide curing system incorporated in it are available, and it is now possible to construct a quantitative formulation. It should be mentioned that at each step of the way, hand drawdowns are made and the material is checked for various adhesive qualities: ease of drying, clarity, temperature needed to laminate, etc.

But the Epon 828 and Versamid 125 are both relatively expensive compared to the cost of acrylic water-dispersions, so using a large amount of the epoxy/polyamide curing additive must be guarded against. If too little is used, the adhesive may not cure enough; if too much is used, the adhesive may be too expensive, and also may be too hard, and hence, may not adhere well. It is at this point that some judicious laboratory experimentation be undertaken.

It will probably be necessary to make the adhesive in two steps, the first being the preparation of the epoxy/polyamide/acid combination with, perhaps, other additives put in. The second step would be making the total mixture of acrylic, acid, epoxy, polyamide, and whatever other additives might be needed, such as defoamer.

Having done all this, it may be found in using the adhesive that a problem arises in the running of the adhesive, and not necessarily with the adhesive as an adhesive. It may be found that shear stability is poor and that the viscosity tends to rise simply because of the mechanical shear put on the product by the act of pumping it into the pan or the turning of the coating roll.

There is no general rule about what to do to effect shear stabilization. In the case of the combination of ingredients mentioned: acrylic polymer, acid, epoxy resin, and polyamide resin, it was found that the addition of a relatively small amount of *Dodecanol* (0.5%) prevented viscosity build-up and ultimate coagulation in the mixture when subjected to shearing stresses.

It may be that the use of a small quantity of a polar/non-polar substance with some tendency towards autophobism might prevent one group of particles of the dispersion from wetting another group, reducing the possibility of viscosity build-up and ultimate coagulation.

It was necessary to develop a test for the effect of shear stress, and check what the addition of various substances did to the composition in this regard. A suitable test procedure was found to be measuring viscosity (a Zahn cup time was found to be a suitable measurement) before and after 30′ of agitation with a high-speed mixer. If no change in viscosity was found, the composition was considered stable.

A useful waterborne adhesive formulated from the materials discussed above is shown in Table 9.2.

For this particular adhesive to run well, application with a gravure head with a smooth roll and without a doctor blade is needed. It also requires corona-discharge treatment of the substrate to which it is applied. It is

TABLE 9.2. A Useful Waterborne Adhesive.

Ingredient	Percent by Weight
Non-ionic acrylic water dispersion, solids 45.5% (Rohm & Haas HA-8 or equivalent polymer is polyethylacrylate)	79.12
Epon 828 or equivalent	2.15
Versamid 125 or equivalent	1.85
Glacial acetic acid	0.86
Dodecanol	0.5
Defoamer (there are many)	0.7
Water from epoxy/polyamide mixture	1.31
Added water	13.51

typical of waterborne adhesives which contain no alcohol that they do not wet most packaging substrates well and have too low a wet strength to run well by gravure.

A SINGLE GENERAL RULE

The above formulations illustrate how developing a composition for a particular purpose might be done, starting from the beginning. In general, a start is made with a film-forming polymer which has fair to good adhesion to the substrate to which it is applied, to which modifying materials are added as required to accomplish the given purpose. There are many polymers from which to choose.

Once the polymer has been chosen, the range of resin additives and plasticizer additives becomes limited because of the requirement that any of these materials added be compatible (i.e., form solid solutions) with the polymer.

As already mentioned, formulating inks, coatings and adhesives is a process of little science and much art. It requires a knowledge of properties of materials and a kind of intuition as to how to make use of them.

And an infinite amount of patience.

CHAPTER 10

The Permeability of Flexible Packaging Materials

A primary goal of the converter is to produce packaging materials with as low a permeability to gases and vapors as is feasible. To some it may seem strange, but all materials, with the exception of glass and metal, are more or less permeable to such things as water vapor, air, organic vapors, essential oils, etc.

Metal foils are not transparent and glass is not flexible.

If a package of corn or potato chips is readily invaded by water vapor, the chips may become soggy. Should air readily enter a package of luncheon meats, their color may change from red to an unappetizing grey; and if oxygen comes into contact with cheese, mold may develop or the fats may become rancid. When a carton of arrowroot crackers is thrown on a 5-gallon can of kerosene in the storage area of a supermarket, the flavor of the crackers may be altered if the wrapper is not a barrier to the passage of kerosene vapor. The barrier properties of the web become as important a set of characteristics as its strength and sealability.

Thus, over the years, a major thrust of many development programs undertaken by the converter and suppliers to the converter has been to enhance the barrier properties of commercial packaging webs.

An exception is the packaging of produce, products which must respire to avoid rotting. A wrapper for lettuce will be highly permeable.

THE MECHANISM OF PERMEATION OF THIN FILMS

Since the converter depends to an overwhelming extent on organic polymers to control permeability, it is the permeability properties of this class of materials that is of primary concern. However, before embarking on proposing a theory of permeability, some nomenclature should be made clear.

The substance through which something is migrating is called a membrane. The substance migrating is called a permeant. When a permeant migrates through a membrane, it does so in three phases:

- Phase 1 – It dissolves in the membrane.
- Phase 2 – It diffuses through the membrane.
- Phase 3 – It evaporates out of the membrane into its new space.

The implied assumption is that migration is taking place from a region of high concentration of permeant to a region of low concentration. The process is called *activated diffusion*.

Under these circumstances, Fick's first law of diffusion applies:

$$m = -D\,dc/dx \tag{10.1}$$

where:

m = rate of permeant flow
D = diffusion coefficient for the permeant membrane pair, i.e., diffusion of that permeant in that membrane
dc/dx = concentration gradient in direction of flow

dc/dx is positive in a direction of increasing concentration. But diffusion takes place from a high concentration to a lower one, hence, the minus sign.

In most experiments, c, the concentration on the high side of the membrane, is generally held constant, and the concentration on the low side of the membrane changes only slightly when it does change and is frequently zero. Then Equation (10.1) integrates into:

$$m = -D\,[c_1 - c_2]/\delta \tag{10.2}$$

where:

c_1 = permeant concentration on the high concentration side
c_2 = permeant concentration on the low concentration side
δ = thickness of the membrane

Since the greatest concern is with substances where the permeant solubility is low, it can be assumed that Henry's law applies and that

$$c = Sp \tag{10.3}$$

where:

S = the solubility coefficient for the permeant/membrane pair
p = the partial pressure of permeant gas (vapor) over the membrane

or

$$c_1 = Sp_1$$

$$c_2 = Sp_2$$

Substituting Equation (10.3) into Equation (10.2) gives:

$$m = -DS[p_1 - p_2]/\delta \qquad (10.4)$$

Rearranging terms and dropping the minus sign yields

$$P = DS = m\delta/[p_1 - p_2] \qquad (10.5)$$

which defines a permeability constant. (P is always positive and there is no longer a concentration gradient. It has been integrated into the actual concentration values. $c_1 - c_2$ is positive. It is thus legitimate to drop the negative sign.)

The permeability constant is equal to the product of the solubility coefficient and the diffusion coefficient of the permeant/membrane pair. Its units are: quantity of permeant transmitted through the membrane per unit time per unit area per unit thickness per unit partial pressure difference.

In the cgs system of units, this becomes STP (standard temperature and pressure—0°C and atmospheric pressure) cc per cm^2 per mm thickness per sec. per cm Hg partial pressure difference.

Units in common usage are:

- for condensible vapors, e.g., water, organic solvents, etc.: gm per hundred in^2 per day per mil of thickness for a difference of pressure from a saturated atmosphere to zero
- for non-condensible gases, e.g., oxygen, carbon dioxide, nitrogen, hydrogen, etc.: STP cc per 100 in^2 per mil of thickness per day per atmosphere pressure difference

THE DEPENDENCE OF PERMEABILITY ON TEMPERATURE

The mechanism of solution/diffusion/evaporation is highly temperature dependent. In many physico-chemical occurrences such as dependence on temperature of chemical reaction equilibrium constants, many kinetic constants, etc., the variation of the function being considered with temperature follows the form of the Arrhenius equation (named for Svante Arrhenius, a great Swedish chemist), to wit: $d \ln F/dT = H/RT^2$, where F is the function under consideration, H the energy change accompanying the process, R the perfect gas constant, and T the absolute temperature in degrees Kelvin. While this seems to follow the form of the Clausius-Clapeyron equation, it does not, since the Clausius-Clapeyron equation is really $dp/dt = H/TV$ where V is the difference between molal volume of the vapor and molal volume of the liquid, and the form $d \ln p/dt = H/RT$ is an approximation based on the fact that the molal volume of the liquid is very small compared to the molal volume of the gas.

The general equation $d \ln F/dT = H/RT^2$ integrates into $\ln F = \ln F_o - H/RT$, or $F = F_o e^{-H/RT}$. Thus $S = S_o e^{-H/RT}$ where H is the heat of solution, and $D = D_o e^{-E/RT}$, where E is the activation energy for the diffusion process per mol of material diffused, a large number. R is the gas constant per mole and T is the absolute temperature in degrees Kelvin.

Since P, the permeability constant for any permeant/membrane pair is given by the product DS, then $P = D_o S_o e^{-[H + E]/RT}$ where $H + E = H_D$, the activation energy of the permeation process.

$$P = P_o e^{-H_D/RT} \tag{10.6}$$

or, converting into logarithms:

$$\ln P = \ln P_o - H_D/RT \tag{10.7}$$

and, since $\log_{10} x = \log_{10} e \ln x$, and since $\log_{10} e = 2.3026$, $\log P = \log P_o - 2.3026 H_D/RT$ or:

$$\log P = a - b/T \tag{10.8}$$

where $a = \log P_o$ and $b = 2.3026 H_D/R$.

The variation of permeability with temperature has been reduced to a rather simple logarithmic relationship. A plot of log permeability vs. the reciprocal of the absolute temperature yields a straight line. The determination of permeability at two different temperatures will give the ability to find permeabilities over a range of temperatures.

Composite Materials

Passage of permeants through structures consisting of two or more membranes combined together may be calculated if the permeability of each individual component is known. Although one component is adjacent to another, it will be assumed that the component A does not influence the permeability of component B with respect to the permeant under consideration. Then $m = P_i p_i / x_i$ where m = mass of permeant migrating through the composite, P_i = the permeability of the ith layer, p_i is the pressure drop across the ith layer x_i is the thickness of the ith layer. If permeabilities are considered to be similar to electrical conductances, and in a composite they are considered to be in series then:

$$1/P = 1/P_1 \cdot x_1 + 1/P_2 \cdot x_2 + \ldots\ldots\ldots + 1/P_n \cdot x_n \qquad (10.9)$$

if there are n components. x is thickness in mils, as permeability is given per mil of thickness.

In this discussion, it has been assumed that the permeability solubility coefficient and diffusion coefficient are all independent of permeant concentration (or partial pressure) over the range of concentrations encountered in flexible packaging applications.

In all calculations using Equation (10.9), it must be remembered that all components, including adhesives, primers, etc., must be included.

THE MEASUREMENT OF PERMEABILITY

There are a number of ways of measuring permeability, most of which are classified as gravimetric or manometric, with the exception of water-vapor permeability, where an infrared detection technique is used. In the gravimetric case, a dish containing permeant is covered by the membrane, whose permeability is to be measured. The membrane is sealed to the dish in such a way that the exposed area is known and the weight loss over time under standard temperature conditions is measured, thus determining how much permeant has escaped through the membrane. This method is used for organic vapors.

The standard method for determining water-vapor permeability was the Southwick method of sealing calcium chloride into such a dish, exposing the dish to 90–100% R.H., and measuring gain in weight. The standard procedure today is described in detail in ASTM (American Society for Testing and Materials) test method F372-73.

The acceptable manometric procedure for non-condensible gases (oxygen, nitrogen, hydrogen, carbon dioxide, hydrogen sulfide, etc.) is described in detail in ASTM test method D1434-82, including derivation of the formula for calculating permeability from the data.

Permeability Data

The units recommended by many people and institutions are SI units [i.e., kilogram, meter, second, the derived units of the Pascal (unit of pressure), and the Newton (unit of force = 1 kg meter/sec.^{2}), etc.]. Since much of the data in the literature and much of the data determined over the years by the author and his co-workers were in grams per hundred in.^{2} per day per mil of thickness per pressure difference between 90−100% relative humidity and zero relative humidity for water-vapor; and STP cc per hundred in.^{2} per day per mil of thickness per atmosphere partial pressure difference for non-condensible gases, rather than go through the conversions, the units have been left in their old form.

There are a great deal of data in the literature. There are also a good deal of data in manufacturers' bulletins and a considerable amount of unpublished data. In presenting this information, no effort has been made to select what might be more accurate from what might be less accurate. It is assumed that all data presented are equally valid and that the measurements were made with equal care. Some of these are presented in Table 10.1.

It can readily be seen from the data that the permeability of plastic films varies over a wide range; from an oxygen permeability of PVdC of 0.6 at a temperature of 25°C to an oxygen permeability of low-density PE of 690.6 at the same temperature, a difference of three orders of magnitude. There is an indication that the more crystalline the polymer, the lower the permeability generally, since a high level of crystallinity indicates a close packing of polymer chains and fewer spaces available for foreign molecules to migrate into. Yet it is well known that the amorphous nylons are better barriers than the crystalline ones. (An exception that proves the rule?)

The not unexpected result that a plasticized polymer is a poorer barrier than an unplasticized one is shown clearly in the case of PVC, where the presence of dioctyl phthalate produces a substantial increase in the permeability to all non-condensible gases (and probably everything else).

Temperature sensitivity of permeability is dramatic. In most instances a small change in temperature produces a large change in permeability. In Table 10.2, the relationship $\log P = a - b/T$ is shown for a number of permeant/membrane pairs.

Table 10.2 illustrates the general sensitivity to temperature of permeability for all plastic films. The variation in permeability with temperature is very different for different materials and the permeability of each membrane, and its variation with temperature, is quite characteristic of the particular membrane.

Figures 10.1 and 10.2 show a graph of permeability of oxygen in Nylon 6 and in PET. Figure 10.1 is the logarithmic line and Figure 10.2 shows the

TABLE 10.1. Permeability of Various Plastic Films to Some Common Permeants.

Film	Temperature		Water Vapor	Oxygen	Nitrogen	Carbon Dioxide
	K	°F				
PE (density = 0.922)	298.1	77	0.74	690.6	129.6	3277.0
	303.1	86		918.5	317.3	4208.0
	310.9	100	1.3			
PE (density = 0.9380)	298.1	77		270.3	82.4	1007.0
	303.1	86		350.7	110.2	1236.0
	310.9	100	0.7			
PE (density = 0.954)	298.1	77		144.0	41.1	585.0
	303.1	86		183.7	55.1	718.0
	310.9	100	0.33			
Polypropylene (cast)	298.1	77	0.2	280.0	50.8	1193.0
	303.1	86		384.0	73.4	1534.0
	310.9	100	0.4			
Polypropylene (oriented)	298.1	77		260.0	42.0	1050.0
	303.1	86				
	310.9	100	0.26			
Polyester (PET)	298.1	77	2.38	6.3	1.49	21.1
	303.1	86		7.5	1.85	25.0
	310.9	100	2.50			

(continued)

TABLE 10.1. (continued).

Film	Temperature		Water Vapor	Oxygen	Nitrogen	Carbon Dioxide
	K	°F				
Nylon 6	298.1	77		4.75	1.16	11.2
	303.1	86		6.3	1.60	14.7
	310.9	100	8.4			
Polyvinylidene chloride (PVdC, Saran)	298.1	77	0.06	0.6	0.10	3.6
	303.1	86		0.9	0.15	5.0
	310.9	100	0.26			
Polymonochlortrifluoro = ethylene (Aclar)	298.1	77		4.89	0.56	25.3
	303.1	86				
	310.9	100	0.03			
Aclar (80% crystalline)	298.1	77		3.08	0.26	5.9
Polyvinylchloride (PVC) rigid	298.1	77		19.5	6.5	165.7
	310.9	100	2.13			
PVC (30 parts dioctyl phthalate per 100 parts polymer)	298.1	77		100.2	28.6	617.9

TABLE 10.2. Variation of Permeability with Temperature.

Membrane	Permeant	Formula
PE (density = 0.922)	water vapor	$\log P = 5.706 - 1739.1/T$
	nitrogen	$\log P = 10.88 - 2540.1/T$
	oxygen	$\log P = 10.34 - 2236.2/T$
	carbon dioxide	$\log P = 9.489 - 1780.3/T$
PE (density = 0.938)	nitrogen	$\log P = 9.566 - 2279.6/T$
	oxygen	$\log P = 9.280 - 2040.8/T$
	carbon dioxide	$\log P = 8.394 - 1606.6/T$
PE (density = 0.954)	nitrogen	$\log P = 9.336 - 2301.3/T$
	oxygen	$\log P = 8.579 - 1910.6/T$
	carbon dioxide	$\log P = 8.158 - 1606.6/T$
Polypropylene (cast)	water vapor	$\log P = 6.664 - 2195.7/T$
	nitrogen	$\log P = 11.41 - 2891.3/T$
	oxygen	$\log P = 10.76 - 2478.3/T$
	carbon dioxide	$\log P = 9.715 - 1978.3/T$
Polyester (PET)	water vapor	$\log P = 0.887 - 152.17/T$
	nitrogen	$\log P = 5.645 - 1630.4/T$
	oxygen	$\log P = 5.467 - 1391.3/T$
	carbon dioxide	$\log P = 5.847 - 1347.8/T$
Nylon 6	nitrogen	$\log P = 8.236 - 2434.8/T$
	oxygen	$\log P = 8.264 - 2260.9/T$
	carbon dioxide	$\log P = 8.126 - 2198.7/T$
PVdC (Saran)	water vapor	$\log P = 10.68 - 3538.1/T$
	nitrogen	$\log P = 11.25 - 3652.2/T$
	oxygen	$\log P = 11.35 - 3456.5/T$
	carbon dioxide	$\log P = 9.524 - 2673.9/T$
Aclar	nitrogen	$\log P = 10.18 - 3108.7/T$
	oxygen	$\log P = 8.860 - 2438.8/T$
	carbon dioxide	$\log P = 10.01 - 2565.2/T$
Aclar (89% crystalline)	nitrogen	$\log P = 8.237 - 2587.0/T$
	oxygen	$\log P = 8.440 - 2369.6/T$
	carbon dioxide	$\log P = 8.867 - 2413.0/T$
PVC (rigid)	nitrogen	$\log P = 1.532 - 214.38/T$
	oxygen	$\log P = 1.957 - 198.67/T$
	carbon dioxide	$\log P = 2.942 - 215.22/T$

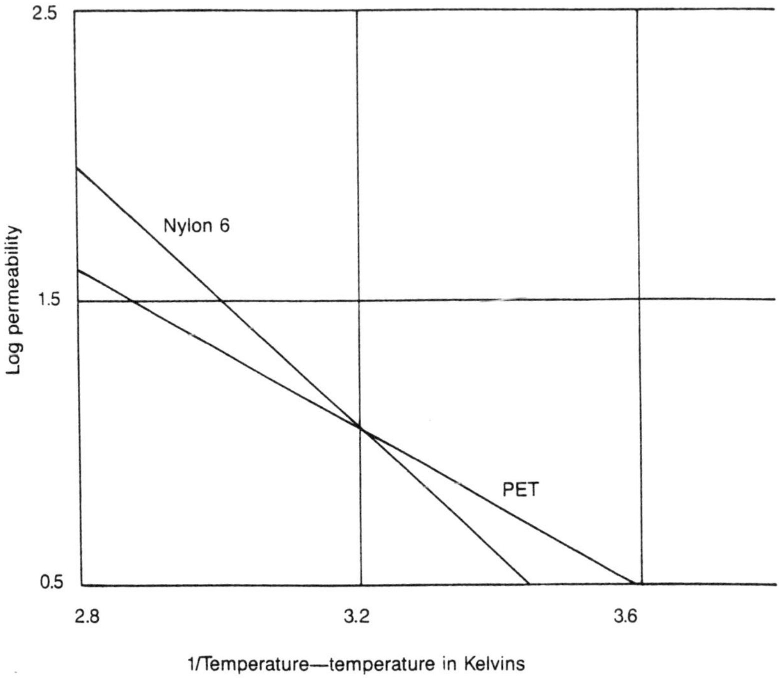

FIGURE 10.1. Log oxygen permeability vs. temperature for Nylon 6 and PET.

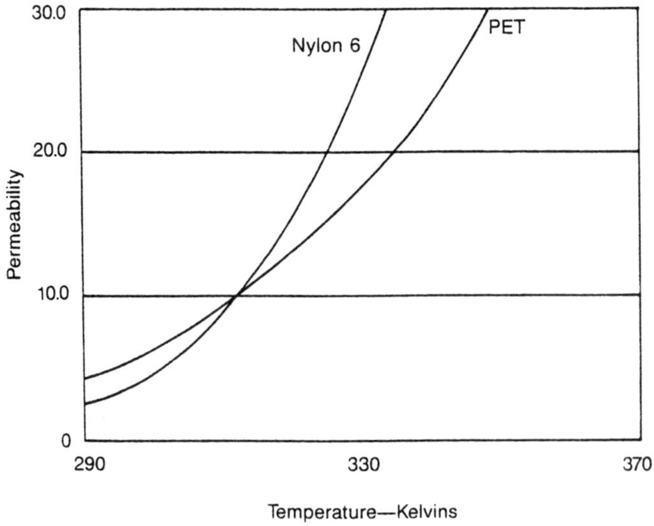

FIGURE 10.2. Oxygen permeability vs. temperature for Nylon 6 and PET.

exponential curves. The curves show quite clearly the difference in temperature variation for the two membranes. At 298.1 K (25°C), the permeability of oxygen in Nylon 6 is 1.16 and in PET is 6.3. The Nylon permeability is about 1/6 that of PET.

At 340 K (67°C), the nylon permeability is 41.1 and for PET it is 23.0 with the nylon permeability being almost double that of the PET. The curves intersect at a temperature of 310.8 K (37.7°C, about 100°F), where the permeabilities of the two membranes are equal (9.85).

Permeability says something about both the permeant and the membrane. For example, the difference between permeability of oxygen and nitrogen, which is consistent from one membrane to the next, speaks of the structural differences between the two molecules.

While the fact that the permeability of carbon dioxide (CO_2) is higher than either of the other two common non-condensible gases may again be explained by the structural differences among the gases, this does not extend to water (H_2O) or hydrogen sulfide (H_2S). Explaining some of the anomalous behavior of permeant/membrane pairs is beyond the scope of this book. Let it suffice to say that it exists.

While cellophane is no longer a major commercial product, there are some interesting data concerning cellophane permeability which merit inclusion in this chapter. They are shown in Table 10.3.

Values of permeability at one level of relative humidity differ from values of permeability at some other level, very likely because of change in water content of the base film. The variation recorded gives minimum and maximum permeability found. In any event, cellophane is quite a good gas barrier, although a poor water-vapor barrier, and coatings are necessary to provide that barrier. What coatings on cellophane seem to do is protect the film from having its properties destroyed because of too high or too low a water content.

TABLE 10.3. Permeability of Cellophane by Various Permeants.

Coating on Film	Relative Humidity	Water Vapor	Nitrogen	Oxygen	Carbon Dioxide
Polymer (PVdC)		0.3–0.5			
Nitrocellulose (N/C)		0.5–1.3			
PE		1.1			
PVdC	0%		0.1–0.3	0.3–0.5	1.0–3.0
	90%		0.8–1.1	0.6–1.4	6.0–9.0
N/C	0%		0.6–1.0	0.6–1.4	2.0–4.0
	90%		0.9–2.7	1.8–4.2	8.0–12.0

Permeation of films by various organic vapors is an interesting study in contrasts, some films being excellent barriers to a broad range of organic compounds and others being no barrier at all. Table 10.4 illustrates this phenomenon.

Some generalization may be drawn from the above data. The polyethylenes, being hydrocarbons themselves, have poor resistance to the passage of hydrocarbon vapors, but are reasonably resistant to the passage of water-soluble materials (e.g., methanol, acetone). The lower the water-solubility, the greater the permeability of the ethylene polymers.

Quite the reverse is true of N/C-coated cellophane. The greater the water-solubility, the higher the permeability; and the lower the water-solubility, the lower the permeability. PVdC-coated cellophane seems impermeable to all organic vapors, regardless of extent of water-solubility. And the polyester film which does not depend on coatings for its general chemical resistance seems all but impermeable to organic vapors. One may expect a high molecular weight, oriented polyester and, perhaps, a polyamide, to be stubborn about allowing gases and vapors easy passage. The chemistry of the polymer, in addition to its structure, seems to play a role in its permeance by many permeants.

There are some data on composites of many varieties made by the converter. Some of this information is shown in Tables 10.5–10.8.

While an undamaged 0.001 metal foil has a zero permeability to water vapor and oxygen, very thin foils have a number of pinholes through which vapors and gases may pass. When the foil is well supported there is little danger of the number of pinholes increasing in handling. Pinholes disappear when the foil reaches a thickness of 1 mil. Below this level, all standard commercial gauges of foil have pinholes. Hence, only the foil laminations containing 1 mil foil (or a greater thickness) will have a zero permeability.

Other, smaller thicknesses, such as 0.5 or 0.35 mil will have small, but non-zero permeabilities. Supporting the foil even with as poor a barrier as PE tends to keep the permeability low, as compared with a thin paper and a non-barrier adhesive.

The data are rather incomplete, but are measurements made and reported. Because there is no knowledge of the permeability of the adhesive layer, or whether the interfaces between two layers cause a sharp change in direction for a diffusing molecule, all calculated values using Equation (10.9) are higher than the measured value. In the case of oxygen permeating 0.5 mil PET/2 mil PE, a calculated value of 11.5 is obtained compared to a measurement of 6.7. This leads to a value of 16 for the permeability of the thin layer of adhesive, a rather doubtful result.

Since it is equally doubtful that Equation (10.9) is so woefully incorrect, the conclusion becomes inescapable that the adhesive at the interfaces with

TABLE 10.4. *Permeability of Various Plastic Films to Sundry Organic Compounds.*

Film	Temperature °C	Compound	Permeability
PE (density = 0.922)	24	heptane	307.0
	23	benzene	513.0
	21	carbon tetrachloride	633.0
	23	methanol	1.3
	24	acetone	20.8
	23	acetic acid	3.9
	24	dibutyl ether	100.0
	24	ethyl acetate	43.0
PE (density = 0.954)	23	heptane	54.0
	23	toluene	120.0
	23	carbon tetrachloride	100.0
	23	methanol	0.22
	23	acetone	3.5
	23	ethyl acetate	4.7
	23	acetic acid	1.0
	23	dibutyl ether	22.0
Cellophane	35	hexane	0.56
(N/C-coated)	35	benzene	0.43
thickness equivalent	35	carbon tetrachloride	0.62
to yield of	35	ethanol	58.0
19,500 in.^{2}/lb.	35	acetane	38.6
	35	ethyl acetate	0.41
	35	acetic acid	39.3
Cellophane	25	ethanol	0.03
(PVdC-coated)	35	ethanol	4.0
thickness equivalent	25	butanol	0.01
to yield of	25	hexane	0.007
19,500 in.^{2}/lb.	35	hexane	0.11
	25	chloroform	0.10
	25	amyl acetate	0.01
	35	benzene	0.16
	35	ethyl acetate	0.17
	35	carbon tetrachloride	0.22
Polyester (PET)	35	hexane	0.09
	35	benzene	0.28
	35	carbon tetrachloride	0.11
	35	acetone	1.3
	35	ethyl acetate	0.0
	35	ethanol	0.0

TABLE 10.5. Permeability of Some Composites Containing Aluminum Foil.

Structure	Water Vapor (100°F)	Oxygen (77°F = 25°C)
20 lb. kraft paper/0.5 mil PE/0.35 mil aluminum foil/1 mil PE	0.013	0.27
30 lb. pouch paper/0.5 mil PE/0.35 mil aluminum foil/1 mil PE	0.013	0.47
60 lb. litho paper/0.5 mil PE/0.35 mil aluminum foil/1 mil PVC	0.013	0.6
19,500 in.^{2}/lb. N/C-coated cellophane/0.35 mil aluminum foil	0.02	
19,500 in.^{2}/lb. N/C-coated cellophane/1 mil aluminum foil	0.0	0.0
19,500 in.^{2}/lb. N/C-coated cellophane/0.5 mil PE/0.35 mil aluminum foil/2 mil PE	0.013	0.8
18 lb. tissue/casein-neoprene adhesive/0.35 mil aluminum foil	0.4	5.0

TABLE 10.6. Permeability of Nylon and Polyester Composites.

Composite	Water Vapor (100°F)	Oxygen (77°F = 25°C)
1 mil Nylon 6/2 mil PE	1.33	1.3
1.5 mil Nylon 6/2 mil PE		1.0
0.5 mil PET/2 mil med. den. PE	0.33	6.7
2 mil PET/1.5 mil Nylon 6	0.7	

TABLE 10.7. Permeability of Paper Composites.

Composite Structure	Water Vapor (100°F)	Oxygen (77°F = 25°C)
25 lb. bleached kraft/1 mil PE	1.5	500.0
30 lb. bleached kraft/1 mil med. den. PE	1.0	
20 lb. glassine/7 lb. wax/20 lb. glassine/0.5 mil PE		1.8
30 lb. pouch paper/0.4 mil PVdC	0.3	5.0
35 lb. glassine/0.4 mil PVdC		1.5
12 point board/2 mil PVdC film	0.2	0.8
30 lb. pouch paper/0.3 mil PVdC/1 mil PE	0.2	1.1
25 lb. glassine/7 lb. wax/25 lb. glassine	0.6	15.0

TABLE 10.8. Permeability of Cellophane Composites.

Composite Structure	Water Vapor (100°F)	Oxygen (77°F = 25°C)
19,500 in.^{2}/lb. K-cello/2 mil PE	0.4	1.3
25,000 in.^{2}/lb. K-cello/0.5 mil PE/25,000 in.^{2}/lb. K-cello	0.4	1.6
0.75 mil OPP/25,000 in.^{2}/lb. K-cello/1.5 mil PE		0.5
1 mil hi. den. PE/19,500 in.^{2}/lb. K-cello/1.5 mil PE		0.3
0.75 mil OPP/25,000 in.^{2}/lb. K-cello/2 mil EVA wax combination		0.7

K-cello = PVdC-coated cellophane. "Cello" is used as an abbreviation for cellophane.

both the PET and the PE plugs the microscopic holes at the surface of at least one of the two films, and slows the passage of gas through the composite. The laminated structure thus consists of the following layers: PET/adhesive interface/adhesive/adhesive interface/PE. The interface assumes the role of a layer in the composite.

Paper composites represent a mixed bag since much depends on the smoothness of the paper when applying a coating. Any sort of barrier properties are very dependent on coating quality, and it is possible this might vary somewhat from one web to another. There may also be considerable variation in the quality of how the measurement of permeability is taken. The data on PVdC-coated and laminated substrates show this. A 2 mil film laminated to a 12 point board has an oxygen permeability scarcely lower than a 0.3 mil coating sandwiched between pouch paper and 1 mil PE. The water vapor transmission rates are identical. Explaining these results would indeed be difficult. But there they are.

Measurements on cellophane composites give results that hang together logically. There is not as much variation in the results from what might be expected.

Measurements of permeability are not of the greatest precision, but do give usable results. When values of permeability reach very low levels, perhaps below 1 gram per 100 in.^{2}/day for condensible vapors or below 1 STP cc per 100 in.^{2}/day for non-condensible gases, it becomes difficult to choose between 0.4 and 0.6 as to which might be the better barrier. At this point, other factors might enter into selecting the superior structure.

Nevertheless, there exists a good deal of understanding of the mechanism of permeation, some reasonably accurate methods of making such measurements, and a rather extensive quantity of data available to the converter.

If the converter understands what his customer's barrier problem is, it is quite possible simply to write down just what the customer may need in the way of a package.

Among the possible sources of information, there are two which are very valuable. One is a chapter in *Engineering Design of Plastics* by C. E. Rogers entitled "Permeability and Chemical Resistance," published by Van Nostrand-Reinhold for the Society of Plastics Engineers. The other is a report by Stannett, Szwarcz, and Myers entitled *Evaluation of Gas Transmission Rates of Flexible Packaging Materials,* submitted to thc U.S. Army Quartermaster Corps under Contract DA 19-129 QM 123, and sold by the Office of Technical Services (OTS) as PB 151124.

Much of the data on composites included here are previously unpublished results determined by the author and his co-workers over a period of many years.

In addition, *Modern Plastics* and *Modern Packaging* (now called *Packaging*) magazines and their annual encyclopedia issues dating as far back as the late 1950s, are valuable sources of information.

CHAPTER 11

The Future of Flexible Packaging

IT is simple enough to wax imaginative when forecasting what might be, 50 or 100 years from now. No one cares. No one remembers what might have been said or written. But if the forecast is for the next decade or two, this must be serious and taken seriously.

The 20th century has been an era of explosive growth in science and technology. From the discovery of X-rays by Roentgen and of the electron by J. J. Thomson, to the mapping of the human genome, from the development of Rayon to the design of software for computers, from Rutherford, Bohr, and Einstein to Hiroshima, the march has gone on, for better or worse. From Bardeen, Brattain, and Shockley to Gates' disc operating system, many possibilities have opened. Instruments like the scanning electron microscopes that have only begun to be used, exist. There has been a steady march of science and technology for the past 100 years. And, without doubt, this will continue.

For better or worse.

Somewhere in this welter of new technical knowledge lie some things that are more important than others, some of which will appear in all their splendor at a later time.

"Coming events cast their shadows before."

Sometimes.

Assuming the aforementioned aphorism, an attempt at some forecasting will be made. As already noted, the upsurge in science and technology will continue unabated. The application of old mathematics and new science will also continue, as in the use of Galois fields in compression of signals for high definition TV. (Evariste Galois was killed in a duel at the age of 20 in 1832.)

Some of this will find its way into converting.

But there is an added wrinkle: environmental regulations.

The Clean Air Act amendments of 1990 require each state to set up permit regulations that will permit factory operations if the factory conforms to certain minimal standards. Should the factory not conform, it will not receive a permit. The EPA has set these minimal standards and minimal penalties for non-compliance. Civil fines may be levied and criminal penalties for a "responsible" individual.

The permit requirement will depend on where the converter is located and the quantity and type of emissions of the plant. For example, ozone is not permitted at all. All corona discharge (CD) units manufacture ozone. All converters operating CD units will be required to install ozone decomposition units. These are commercially available.

The Flexible Packaging Association, in conjunction with its legal counsel, should draw up a list of guidelines for its members, if it has not already done so.

But each converting company should take responsibility for itself. Regardless of state, it ought to set up a department to deal with environmental problems, even if the department consists of a single individual. It also might consider hiring a consultant, of which there are many, some of whom are quite competent.

The time frame is a little vague, but certainly each state will have its permit procedures set up within the next two years.

Obviously, the citizenry wants some control over the quality of the air it breathes and it is advisable that the converter comply with what will have the force of law.

If the converter pays attention to EPA regulations, coping with them will not be a problem. If they are ignored or fought, the result may be disastrous.

Some of the regulations may be contradictory, and some might seem unreasonable. Patience and persistence will help.

MATERIALS

It is doubtful if any major new films will be developed. There are so many now, and they have not been fully exploited.

There may be some significance to the vacuum deposition of glass now being explored experimentally. There may also be some significance in the synthesis of two-dimensional polymers, which have not yet come out of the laboratory.

But in the area of coatings and inks and adhesives, there are a number of distinct new possibilities. The uses of 100% solids materials are far from over. They have hardly begun.

Polyol-isocyanate prepolymers, polyamine-isocyanate prepolymers, mixed together and heated, or mixed together and exposed to some form of radiation, ranging from radio waves to electron beams, is one way to go. Or, perhaps, a polymer dissolved in a monomer, properly catalyzed and exposed to radiation. Solvents, with or without incineration, will become obsolete. Pressure from environmentalists will cause this obsolescence.

Concomitant with this will occur further development of waterborne formulations incorporating a variety of catalysts, including some enzyme-like types or perhaps actual enzymes. This technology has developed enormously in recent years. It is invading industrial processes.

It is conceivable that formulations might be made where linkage of adhesive to substrate might be catalytically produced.

And, it is also conceivable that non-polyolefinic coextrusions that adhere well might be made by exposure of the components to radiation in the spaces between individual films before combining, or by preforming a film, activating it and extruding a second film on the first, not truly a coextrusion, but the result will be the same.

EQUIPMENT

There will probably be an increased use of radiation for drying (microwave) and curing of coatings, inks, adhesives. Militating against this development is the need for safety, and hence the large expense of such equipment. But as resistance to solvent emissions grows and it is discovered that formulating in water has its limitations, pressure will build for the use of radiation.

Then again, it may also be found that by the use of enzymes, there is no need for radiation.

Perhaps.

However, there is little doubt that machinery will be built fully instrumented, programmable, perhaps robot-controlled.

There will be, among others, social and economic forces driving the industry. One will be the struggle to clean up the environment. Another will be the much talked of global economy. The prospect of reduction in manufacturing unemployment may be questionable. In order for all industry to compete globally, it will be necessary for all industry to automate as far as possible to reduce the labor component of cost as much as possible and to raise the quality component of the product as high as possible.

It is not too likely that this trend will reach the converting industry as a totality, but part of it will. Competition from Mexico and Canada is to be expected, forcing some automation in the U.S. industry.

It is also conceivable that laminations will be made without an adhesive, as 3M makes Scotchpak, by exposing two surfaces of two disparate films to the proper radiation and bonding the two films together, exposed surface to exposed surface. A corn chip package of two exterior PVdC-coated OPP films bonded together on the uncoated sides might make a very inexpensive snack package.

If we wish to indulge in flights of fancy, we can imagine bacteria or single-celled animals created by biotechnology that feed on the plastic films the converter uses, thus making the converter's waste biodegradable. And these single-celled animals and plants will do no harm elsewhere. (Hah!)

Maybe we should recognize that making PE and other such materials biodegradable is not in the cards for the near term. Maybe not every problem can be solved. Maybe some must be avoided.

What does the future look like? A mixed bag, like the past and the present. There will be new technology and, initially, it will be profitable for those using it. Then it will spread and all, or nearly all, will use it. Price-cutting will cause profit margins to fall, leading to the introduction of another new technology that will make the old product cheaper and better or introducing new products that are cheaper and/or better than the old ones. It will spread, etc., etc., ad infinitum, ad nauseam.

There will be no miracles.

But there will be new technology.

BIBLIOGRAPHY

Advances in Chemistry Series #43. 1964. *Contact Angle, Wettability and Adhesion.* American Chemical Society, Washington, DC.

Bird, R. B., W. E. Stewart and E. N. Lightfoot. 1960. *Transport Phenomena.* John Wiley & Sons, New York.

Carslaw, H. S. and J. E. Jaeger. 1959. *The Conduction of Heat in Solids, 2nd Edition.* Oxford University Press, New York.

Carslaw, H. S. and J. E. Jaeger. 1963. *Operational Methods in Applied Mathematics.* Dover Publications, New York.

Eschbach, O. W. and M. Souders. 1975. *Handbook of Engineering Fundamentals, 3rd Edition.* John Wiley & Sons, New York.

Cotton, J. W., Chairman, Editorial Committee. 1980. *Flexography—Principles and Practices, 3rd. Edition.* Flexographic Technical Association, Ronkonkoma, New York.

Jakob, M. 1949, 1953. *Heat Transfer.* John Wiley & Sons, New York.

Kreyszig, E. 1962. *Advanced Engineering Mathematics.* John Wiley & Sons, New York.

McKelvey, J. M. 1959. "Fundamentals," in *Processing of Thermoplastic Materials,* E. C. Bernhardt, ed., Van Nostrand Reinhold, New York.

Rogers, C. E. 1964. "Permeability and Chemical Resistance," in *Engineering Design of Plastics,* E. Baer, ed., Van Nostrand Reinhold, New York.

Schlichting, H. 1955. *Boundary Layer Theory.* McGraw-Hill, New York.

Spiegel, M. R. 1990. *Theory and Problems of Laplace Transforms, 22nd Printing.* Schaum Outline Series, McGraw-Hill, New York.

Stannett, V. S., M. Szwarcz and A. W. Myers. Undated. *Evaluation of Gas Transmission Rates of Flexible Packaging Materials.* U.S. Government Contract DA 19-129 QM 123. Sold by the Office of Technical Services (OTS) as PB 151124.

Also series of volumes published for the Society of Plastic Engineers by Van Nostrand Reinhold, New York, called *Polymer Science and Engineering.*

Articles published in the trade press since the late 1950s. These publications include: *Modern Packaging,* now called *Packaging; Modern Plastics; Polymer Science and Engineering; Paper, Film and Foil Converter;* and *Food and Drug Packaging.*

INDEX